THE QUEST FOR A UNIVERSAL THEORY OF LIFE
Searching for Life As We Don't Know It

Integrating both scientific and philosophical perspectives, this book provides an informed analysis of the challenges of formulating a universal theory of life. Among the issues discussed are crucial differences between definitions and scientific theories and, in the context of examples from the history of science, how successful general theories develop. The central problem discussed is two-fold: First, our understanding of life is still tacitly wedded to an antiquated Aristotelian framework for biology; second, there are compelling reasons for considering that familiar Earth life, which descends from a last universal common ancestor, is unrepresentative. What is needed are examples of life as we don't know it. Potential sources are evaluated, including artificial life, extraterrestrial life, and a shadow biosphere right here on Earth. A novel strategy for searching for unfamiliar life in the absence of a definition or general theory is developed. This book is a valuable resource for graduate students and researchers studying the nature, origins, and extent of life in the universe.

CAROL E. CLELAND is Professor of Philosophy at the University of Colorado, Boulder, USA, where she is also Director of the Center for the Study of Origins and a Co-Investigator at the Center for Astrobiology. She teaches advanced interdisciplinary courses in the philosophy of science, including graduate seminars on central issues in astrobiology. She publishes papers in major science and philosophy journals, and is co-editor of the anthology *The Nature of Life* (Cambridge University Press, 2010).

"An essential read for anyone interested in the nature of life and its origins. Cleland's philosophical outlook means that she approaches the subject from a fresh perspective, framing important questions rarely discussed by scientists ... and deliberating points in a provocative way that forces readers to examine some of their cherished beliefs that they thought were self-evident."

Athel Cornish-Bowden,
French National Center for Scientific Research, Marseilles, France

"What is life? What universal principles apply to any biosphere? Our efforts to answer these deep questions are stymied because of our biased, Earth-bound perspective with only one kind of (known) life. In a book rich with original ideas and lucid insights, science philosopher Carol Cleland considers life from the perspective of what we *don't* know – the limitations, hidden biases, sloppy definitions, and false assumptions that may lead us astray. From 'shadow biospheres' on Earth, to artificial life in the lab, to alien lifeforms in distant galaxies, Cleland expands our minds and leads us to rethink what we thought we knew."

Robert Hazen,
Carnegie Institution for Science, Washington, USA

"Searching for life elsewhere in our solar system or beyond is at the forefront of science today due to recent discoveries about terrestrial life, planetary environments, and planets around other stars. We can't extrapolate from our single example of life on Earth, which all share common biochemistry and are descended from a common ancestor, as to what the characteristics of life elsewhere in the universe might be. Given these uncertainties, how can we identify something as alive? What does it mean to be living? What is life? Carol Cleland takes a philosophy of science approach to what constitutes life, integrating it with biology in a planetary context. She has made a valuable contribution to our understanding of the nature of life and how to search for it, both on Earth and elsewhere."

Bruce Jakosky,
University of Colorado, Boulder, USA

CAMBRIDGE ASTROBIOLOGY

Series Editors

Bruce Jakosky, Alan Boss, Frances Westall, Daniel Prieur, and Charles Cockell

Books in the Series:

1. *Planet Formation: Theory, Observations, and Experiments*
   Edited by Hubert Klahr and Wolfgang Brandner
   ISBN 978-0-521-18074-0

2. *Fitness of the Cosmos for Life: Biochemistry and Fine-Tuning*
   Edited by John D. Barrow, Simon Conway Morris, Stephen J. Freeland, Charles L. Harper, Jr
   ISBN 978-0-521-87102-0

3. *Planetary Systems and the Origin of Life*
   Edited by Ralph Pudritz, Paul Higgs, and Jonathan Stone
   ISBN 978-0-521-87548-6

4. *Exploring the Origin, Extent, and Future of Life: Philosophical, Ethical, and Theological Perspectives*
   Edited by Constance M. Bertka
   ISBN 978-0-521-86363-6

5. *Life in Antarctic Deserts and Other Cold Dry Environments*
   Edited by Peter T. Doran, W. Berry Lyons, and Diane M. McKnight
   ISBN 978-0521-88919-3

6. *Origins and Evolution of Life: An Astrobiological Perspective*
   Edited by Muriel Gargaud, Purificación Lopez-Garcia, and Hervé Martin
   ISBN 978 0521-76131-4

7. *The Astrobiogical Landscape: Philosophical Foundations of the Study of Cosmic Life*
   Milan M. Ćirković
   ISBN 978 0521-19775-5

8. *The Drake Equation: Estimating the Prevalence of Extraterrestrial Life through the Ages*
   Edited by Douglas A. Vakoch and Matthew F. Dowd
   ISBN 978-1-107-07365-4

9. *Astrobiology, Discovery, and Societal Impact*
   Steven J. Dick
   ISBN 978-1-108-42676-3

10. *Solving Fermi's Paradox*
    Duncan H. Forgan
    ISBN 978-1-107-16365-2

11. *The Quest for a Universal Theory of Life: Searching for Life As We Don't Know It*
    Carol E. Cleland
    ISBN 978-0-521-87324-6

THE QUEST FOR A UNIVERSAL THEORY OF LIFE

Searching for Life As We Don't Know It

CAROL E. CLELAND

University of Colorado, Boulder

CAMBRIDGE
UNIVERSITY PRESS

CAMBRIDGE
UNIVERSITY PRESS

Shaftesbury Road, Cambridge CB2 8EA, United Kingdom

One Liberty Plaza, 20th Floor, New York, NY 10006, USA

477 Williamstown Road, Port Melbourne, VIC 3207, Australia

314–321, 3rd Floor, Plot 3, Splendor Forum, Jasola District Centre, New Delhi – 110025, India

103 Penang Road, #05–06/07, Visioncrest Commercial, Singapore 238467

Cambridge University Press is part of Cambridge University Press & Assessment,
a department of the University of Cambridge.

We share the University's mission to contribute to society through the pursuit of
education, learning and research at the highest international levels of excellence.

www.cambridge.org
Information on this title: www.cambridge.org/9780521873246

DOI: 10.1017/9781139046893

© Carol E. Cleland 2019

First published 2019

A catalogue record for this publication is available from the British Library

Library of Congress Cataloging-in-Publication data
Names: Cleland, Carol, author.
Title: The quest for a universal theory of life : searching for life as we don't know it /
Carol E. Cleland, University of Colorado, Boulder.
Description: Cambridge, United Kingdom ; New York, NY : Cambridge University Press, 2020. |
Series: Cambridge astrobiology | Includes bibliographical references and index.
Identifiers: LCCN 2019010797 | ISBN 9780521873246 (hardback : alk. paper)
Subjects: LCSH: Life–Origin–Philosophy. | Life (Biology)–Philosophy.
Classification: LCC QH325 .C54 2020 | DDC 576.8/3–dc23
LC record available at https://lccn.loc.gov/2019010797

ISBN   978-0-521-87324-6   Hardback

For Leta and Elspeth

Contents

Acknowledgments   *page* xii

Introduction   1

1 The Enduring Legacy of Aristotle: The Battle over Life as
   Self-Organization or (Genetic-Based) Reproduction   8
   1.1 Overview   8
   1.2 Aristotle on the Nature of Life: Nutrition Versus Reproduction   10
   1.3 Classical Mechanism about Life: From Optimism to Quiet
        Desperation   15
   1.4 Darwin to the Rescue?   18
   1.5 Here We Go Again: Aristotelian Roots of Contemporary
        Accounts of the Nature and Origin(s) of Life   21
        1.5.1 Metabolism-Based Versus Evolution-Based Definitions
               of the Nature of Life   22
        1.5.2 Metabolism-First Versus Genes-First Theories
               of the Origin(s) of Life   25
   1.6 Concluding Thoughts   29

2 Why Life Cannot Be Defined   33
   2.1 Overview   33
   2.2 Popular Definitions of Life   36
        2.2.1 Thermodynamic Definitions   36
        2.2.2 Metabolic Definitions   37
        2.2.3 Evolutionary Definitions   40
        2.2.4 Defining Life as Self-Organized Complexity   43
   2.3 The Problem with Definitions   45
        2.3.1 Logical Character of Definition   45
        2.3.2 Limits of Definition   50

2.3.3   Diagnosing the Problem: A Defective Theory of Meaning
        and Reference   53
2.3.4   Why Natural Kinds Cannot Be "Defined"   55
2.3.5   Is life an Exception to the Rule?   59
2.4   Concluding Thoughts   61

3   What Is a Scientific Theory?   63
3.1   Overview   63
3.2   The Syntactic Conception of Scientific Theories   65
3.3   The Semantic Conception of Scientific Theories   68
3.4   Scientific Theories and Definitions   73
3.4.1   Scientific Theories Do Not "Define" Natural Kinds   74
3.4.2   Nonstandard Definitions Do Not "Define" Scientific
        Theories   76
3.5   Concluding Thoughts   79

4   How Scientific Theories Develop   82
4.1   Overview   82
4.2   How Scientifically Fruitful Ontologies Develop: Content Matters   84
4.3   The Goldilocks Level of Abstraction   88
4.4   The Threat Posed by Premature Commitment to Ontologies   93
4.5   The Monist (Versus Pluralist) Stance   98
4.6   Concluding Thoughts   102

5   Challenges for a Universal Theory of Life   105
5.1   Overview   105
5.2   The Magnitude of the $N = 1$ Problem of Biology   107
5.3   Microbes: The Most Representative and Least Well Understood
        Form of Earth Life   114
5.3.1   Planet of the Microbes   115
5.3.2   A Brief History of Misunderstandings and Surprises   118
5.4   The Problem of Contingencies and the Origin(s) of Life   120
5.4.1   A Plague of Contingencies (on Both the SM World and the
        RNA World)   123
5.4.2   The Origin Versus Nature Problem   127
5.5   Concluding Thoughts   130

6   Rethinking the Traditional Paradigm for Life: Lessons from the World
        of Microbes   132
6.1   Overview   132
6.2   Evolution Viewed Through the Lens of the Microbial World   133
6.2.1   The Concept of a Biological Species and the Tree of Life   134

6.2.2   Is Lamarck Hiding in the Shadows?   138
6.3   The Living Individual Viewed Through the Lens of the
Microbial World   143
6.3.1   Is the Host–Microbiome Complex (Holobiont)
a Living Thing?   146
6.3.2   Biofilms: Aggregates of Cells or Living Individuals?   149
6.3.3   Could Rock-Powered Ecosystems Be Living Things?   155
6.4   Concluding Thoughts   159

7   Artificial Life: Could ALife Solve the $N = 1$ Problem?   161
7.1   Overview   161
7.2   Soft ALife: Digital Organisms?   161
7.3   Hard ALife: Living Robots?   164
7.4   Synthetic Biology: Creating Novel Life in the Laboratory?   167
7.5   Concluding Thoughts   170

8   Searching for Extraterrestrial Life Without a Definition or Universal
Theory of Life   172
8.1   Overview   172
8.2   A Case Study: The Viking Missions to Mars   172
8.3   The Role of Anomalies in Scientific Discovery   176
8.4   Searching for Anomalies Using Tentative (Versus Defining)
Criteria   184
8.5   Concluding Thoughts   193

9   A Shadow Biosphere: Alien Microbes on Earth?   195
9.1   Overview   195
9.2   How Scientifically Plausible Is a Shadow Biosphere?   196
9.2.1   Did Life Originate Only Once on Earth?   197
9.2.2   Could the Present-Day Earth Be Host to a Shadow
Biosphere?   201
9.3   If They Exist, Why Have We Not Found Them?   206
9.3.1   Limitations to Microscopy   206
9.3.2   Limitations to Cultivation   207
9.3.3   Limitations to Metagenomic Methods   208
9.4   Potentially Biological Anomalies: Have We Already
Encountered Them?   211
9.5   Concluding Thoughts   215

Conclusion   217

References   220
Index   243

Acknowledgments

This book is an ambitious project. It has been in the works for a long time. During this time, I have had extensive and very helpful conversations with many friends and colleagues about issues covered in this book. I am very grateful to you all.

I would like to single out a few of you for special thanks. I cannot thank Athel Cornish-Bowden and Heather Demarest enough for reading through most of the chapters of this book and providing me with constructive criticism. Thanks to Mitzi Lee for her patience with my many questions and concerns about Aristotle's biology; she saved me from several grave errors of interpretation. I am also grateful to Alexandra Laird, who engaged in the arduous task of putting together the references and index for the book. I am very grateful to Norm Pace for long phone conversations over the years about microbiology. Thanks to Lisa Lloyd for her enthusiasm for the project and lively discussions during our delightful "philosophical breakfasts" in the summers of 2014 and 2015. I would also like to thank Mark Bedau, Nick Byrd, Graeme Forbes, Jim Griesemer, Elizabeth Griffiths, Bruce Jakosky, Mathew Kopec, Maureen O'Malley, Jessica Riskin, Julia Staffel, Mike Stuart, Brian Zaharatos, and Michael Zerella for reading individual chapters and providing me with thought provoking criticisms; their feedback has made the book much better than it would otherwise be.

I have given many talks on issues in this book to different groups of scientists and philosophers and would like to thank them all for helpful discussions and comments. I am especially grateful to John Norton and my fellow visiting scholars at the Center for Philosophy of Science (University of Pittsburgh) for extended discussions about artificial life, definitions, and the search for extraterrestrial life while I was on Sabbatical Leave during the spring of 2016. I also greatly benefited from discussions about natural kinds, definitions, and scientific theories with Lisa Lloyd's energetic philosophy of biology reading group at Indiana University during my visit in 2015. Last but certainly not least, I owe an important debt to the up and coming philosophers (Alexander Beard, Christopher Dengler,

Alexandra Laird, Lorenzo Nericcio, Cristian Larroulet Philippi, and Erich Riesen) and scientists (Ellie Hara, Cara Lauria, and Michael Zawaski) in my Fall 2017 graduate seminar on life who enthusiastically worked through the penultimate version of this book, chapter by chapter, providing me with extensive and helpful feedback. I would also like to thank Eric Parrish for assisting with the figures in Chapter 5.

I am grateful to the NASA Astrobiology Institute (NAI) for providing partial funding (through a grant to the University of Colorado's Astrobiology Center) for the early stages of this project. Thanks also to Sarah Warner who pressured me into getting started on what seemed an almost impossible task, and to the editors at Cambridge University Press for their patience, understanding, and encouragement.

Introduction

This book focuses on the search for a universal theory of life. It is concerned with the history of attempts to develop such a theory, diagnosing why these efforts have thus far been unsuccessful, and determining what is required to forge ahead and successfully pursue such a theory. It is of course possible that the diverse phenomena of life lack an objective natural unity, and hence that no such theory will ever be forthcoming. Indeed, this view has become popular among some biologists and many philosophers of biology. One of the central themes of the book is that skepticism about the prospects of universal biology is not only very premature but also potentially self-fulfilling: One does not want to short-circuit the potentially successful pursuit of universal biology by rejecting it out of hand.

This book is designed for both scientists and philosophers. Its goal is to integrate pertinent philosophical material that is unfamiliar to most scientists with scientific material (from microbiology, biogeochemistry, and planetary science) that is not well known to many philosophers. It should thus be accessible to the well-educated reader who is a specialist in neither area. Nevertheless, scientifically inclined readers may find some of the philosophical discussions tedious and difficult; Chapters 2 and 3 are challenging for anyone without a background in logic and technical philosophy. Similarly, philosophically inclined readers may struggle a bit with some of the scientific material. While it may be tempting to skip over areas with which one is least familiar, I urge the reader to refrain from doing so. To fully understand the challenges posed by the pursuit of a truly general theory of life, readers need to gain the broader interdisciplinary perspective provided by this book. In this context, a brief thematic roadmap, providing an overview of the philosophical and scientific issues covered in this book and how they relate to each other, may be helpful.

Chapter 1 traces the pursuit of universal biology back more than two thousand years to its roots in the work of the Greek philosopher Aristotle, who is also credited with being the first biologist. Aristotle's lasting influence on biology, which is not widely appreciated, is readily seen in the following characteristics, which are held up by modern biologists as fundamental to life: (1) the capacity to self-organize and maintain self-organization for an extended period of time against both external and internal perturbations (metabolism) and (2) the capacity to reproduce and (in light of Darwin's theory of evolution) transmit to progeny adaptive characteristics. I will henceforth refer to the former as "O" and the latter as "R." As Chapter 1 discusses, the conceptual parallels between characteristics O and R and Aristotle's ideas about life are remarkably close. Aristotle identified "nutrition" (metabolism) and "reproduction" as the basic functions of life, and debated (as do so many contemporary researchers) which is more fundamental. Aristotle also bequeathed to biology the thorny problem of teleology, that is, the notion that the basic functions of life involve a strange form of causation that is intrinsically directed at achieving a future goal. On Aristotle's account, organisms are not just fed, they feed *themselves*, and they are not just copied, they copy *themselves*. Chapter 1 traces the history of efforts to exorcise the problematic notion of goal-directed self-causation from biology. Darwin is often credited with having done so but, as will become apparent, it still lurks at the foundations of biology in popular definitions of life and models of the origin(s) of life, both of which are closely patterned on characteristic O or R.

Despite strenuous efforts over the past couple of hundred years, biologists have yet to come up with an empirically fruitful, truly general theory of familiar Earth life, let alone one that applies to all life, wherever and whenever it may be found in the universe. Some philosophers and scientists have responded by giving up on the program of universal biology. The great eighteenth century philosopher and astronomer, Immanuel Kant, despairingly concluded that there would never be a Newton of biology. Many contemporary biologists delight in showcasing exceptions to any allegedly universal principle of life. As a result, most philosophers of biology now embrace a pluralist view of biology, contending that, unlike other natural categories (e.g., water) studied by modern science, life does not have a unified nature. There is another, rarely entertained, possibility, however: The failure of biologists to come up with a universal theory of life is at least in part the result of tacit commitment to an anachronistic Aristotelian framework for reasoning about life. The latter is a central theme of this book.

This issue is taken up again in Chapter 4, which explores, in the context of examples from the history of science, factors that can thwart the development of successful scientific theories. One of the most serious of these factors is commitment to a defective theoretical framework of core concepts for reasoning about a domain of natural phenomena. Pick the wrong theoretical concepts and empirically fruitful general principles will not be forthcoming. Notions of impetus (active internal principles of motion) and rest (the natural state of an inanimate terrestrial body), inspired by Aristotle's writings on motion, held a grip on the scientific mind until around the sixteenth century. Newton's universal theory of motion – consisting of three laws of terrestrial motion and a law of universal gravitation – critically depended upon replacing these ideas with the very different concept of inertia (passive resistance to a change in motion). Similarly, the development of a universal (chemical) theory of material substance occurred rapidly after Lavoisier's discovery that Aristotle was wrong in thinking that water (along with air, earth, and fire) was a basic element. Viewed from this perspective, it seems somewhat scandalous that biology is the only natural science still dominated (albeit often tacitly) by Aristotelian ideas.

Returning to Chapter 2, another central theme of this book is that the popular scientific project of defining life is profoundly mistaken. Definitions of life, which are invariably founded upon characteristics O and/or R, dominate scientific discussions about the nature of life. The most informative definitions supply necessary and sufficient conditions for membership in the class of living things, which has the advantage of exhaustively subdividing the world into things that are living and things that are not. An astrobiologist could (at least in principle) determine whether an extraterrestrial system on Mars or Titan is a living thing simply by applying whatever definition of life they have agreed upon. Similarly, ALife (artificial life) researchers could invent new forms of life simply by tailoring their inventions to such a definition. As discussed in Chapter 2, however, a definition of life is more likely to hinder than facilitate the discovery of truly novel life.

The problem is that definitions are concerned with language and human concepts – with the analysis of the meanings (*qua* concepts in our heads) of terms (words and expressions). Defining bachelor as an unmarried human male provides a good illustration. It tells us what the word 'bachelor' means by dissecting our concept of bachelor into other concepts (being unmarried, human, and male); anything that is unmarried, human, and male is *by definition* a bachelor and vice versa. As Chapter 2 discusses, however, things are not quite this simple. Natural languages are vague. On our definition of 'bachelor', a male infant would qualify as a bachelor, and attempts to fix this problem by adding 'adult' will not fully resolve it. Still, ignoring the problem of vagueness, our definition of bachelor provides a pretty good answer to the question 'what is a bachelor?'

The question is, could a definition of 'life' provide an equally satisfactory answer to the question 'what is life?' Unfortunately, the answer is 'no'. When a scientist asks 'what is life?' she is not interested in the meaning of the word 'life'. She wants to know what life *truly* is: What distinguish bacteria, slime molds, fungi, fish, trees, elephants, and all other living things, wherever and whenever they may be found, from nonliving physicochemical systems. Analysis of the twenty-first century concept of life is unlikely to provide her with a scientifically satisfying answer to this question.

So why does a definition of 'bachelor' provide a good answer to the question 'what is a bachelor?' but a definition of 'life' does not provide a good answer to the question 'what is life?' The answer is complex, having to do with the logical character of definition and an important distinction in philosophy of language between terms designating categories (e.g., water and star) that would exist had there been no human beings and categories (e.g., bachelor and garbage) that are carved out by human interests and concerns; philosophers dub the former 'natural kinds' and the latter 'non-natural [human] kinds'. To make a long story short, if life (like water) is a natural kind, attempts to "define" it will be futile; the extensions of natural kind terms are too open ended to be decided by means of necessary and sufficient conditions. On the other hand, if life is not a natural kind, the scientific program of universal biology is doomed; for (as pluralists contend) the diverse phenomena of life have no natural unity. Either way, as Chapter 2 explains, life cannot be "defined."

Some scientists and philosophers who advance "definitions" of life do not have in mind the traditional logical notion of definition. This is the topic of Chapter 3. Nonstandard definitions resemble traditional definitions structurally in supplying necessary and sufficient conditions for membership in a natural kind. Their authority derives from empirical investigations, however, as opposed to analyses of human concepts. In this way philosophers advocating nonstandard definitions of life hope to evade the serious logical problems discussed in Chapter 2; most scientists are not familiar with those difficulties.

Defenders of nonstandard definitions of life are keen to establish a close logical relationship between the structures of definitions and scientific theories, contending either that (i) a scientific theory of life can be somehow compressed into a statement supplying necessary and sufficient conditions for life or that (ii) a statement supplying necessary and sufficient conditions for life can be logically inferred from a scientific theory of life; on the latter view, a nonstandard definition of life could be viewed as a crucial building block for a theory of life. In this context, Chapter 3 begins with an extended discussion of scientific theories, paying close attention to the highly influential formal (syntactic and semantic) conceptions, which reconstruct scientific theories along the lines of those of mathematics. As will become clear, while both the syntactic and semantic conceptions of scientific theory support the claim that a theory of life can be compressed into a definition-like statement, neither supports the narrower claim that supplying necessary and sufficient conditions *for life* (*qua* natural kind) can encapsulate a scientific theory of life. Moreover, neither conception of scientific theory supports the claim that one can logically infer necessary and sufficient conditions for life from a scientific theory of life. The arguments in this chapter are somewhat technical and difficult but well worth slogging through. What they establish is that scientific theories are far more open ended and fluid than mathematical theories; otherwise they could not accommodate novel and sometimes startling empirical discoveries, which are ultimately a major source of the empirical fruitfulness of a scientific theory. Viewed in this light it is not surprising that formal conceptions of scientific theory have fallen out of favor among philosophers of science. They have been replaced by informal, model-based approaches, which provide no support for the claim that nonstandard definitions can provide a scientifically compelling answer to the question 'what is life?' It is high time that scientists and philosophers abandon the project of defining life in all its nefarious guises.

As alluded to earlier, Chapter 4 focuses on conditions that facilitate and hinder the development of successful scientific theories. This focus is important for the program of universal biology because we currently lack a truly general theory of life. At best, given our limited experience with life (viz., familiar Earth life), we are in the early stage of formulating such a theory. As Chapter 4 explains, the development of an empirically fruitful scientific theory critically depends upon the selection of a promising ontology (set of basic theoretical concepts) for generalizing about a domain of natural phenomena. There are many different, seemingly natural, ways of carving up a domain of natural phenomena into basic categories, most of which are incapable of supporting scientifically fruitful generalizations.[1] Newton's predecessors failed to come up with a truly general theory of motion because they were working with unpropitious theoretical concepts such as impetus, bulk (occupied volume), impenetrability, and weight. Newton succeeded because he opted for (what turned out to be) a more promising concept (inertia). Unfortunately, as Chapter 4 discusses, when faced with predictive or explanatory failure, scientists have a tendency to retain ontologies and revise the generalizations based upon them.[2] This flawed approach can significantly delay the development of a successful, truly unifying, scientific theory. It is thus important to resist becoming prematurely wedded to a theoretical framework for generalizing about a domain of natural phenomena.

---

[1] As this chapter discusses, scientific pluralists are thus right in maintaining that th... ...re equally legitimate ways of carving up a domain of natural phenomena into categories *depending upon one's interests and concerns*. But as Chapter 4 explains, most ontologies will not yield empirically and theoretically powerful generalizations *of the sort sought by scientists*; indeed, the history of science suggests that those that will are few and far between. Instead of taking a stand on the contentious issue of whether monism or pluralism is true of the natural world – an issue that is difficult to defend scientifically or philosophically – I defend (what I dub) a "monist stance" towards natural phenomena, most notably life, over Kellert and colleagues' (2006, xiii–xvi) pluralist stance.

[2] As Chapters 4 and 6 discuss, examples from biology are not hard to find. The bacterial species problem provides an especially salient illustration. In the face of unsuccessful efforts to extend a eukaryote-centric concept of species to unicellular microorganisms, some scientists and many philosophers of biology propose a pluralist conception of species, in effect retaining it (come what may) instead of seeking newer, more unifying concepts for understanding the evolutionary relatedness of known life on Earth.

The tendency to retain ontologies in the face of predictive and explanatory failure is a serious concern for the program of universal biology. As Chapter 5 discusses, compelling theoretical and empirical reasons exist for suspecting that familiar Earth life may not be representative of life considered generally. Molecular biologists have established that all known life on Earth today descends from a last universal common ancestor (LUCA), which means that our experience with life is limited to a single example; this is known as the $N = 1$ problem of biology. Moreover, biochemists have established that life could be at least modestly different at the molecular and biochemical level, and they concede that it is not at all clear how different it could be. Last but certainly not least, since the time of Aristotle, most theorizing about life has been based upon complex multicellular eukaryotes (primarily plants and animals), which are highly specialized, fragile latecomers to our planet. We now know that archaea and bacteria are far older and more diverse, both genetically and metabolically, than multicellular eukaryotes. They are also the most common form of life on Earth today. The human body, for instance, contains many more archaea and bacteria than somatic cells, and the number of viruses in the human body far outnumbers those of archaea and bacteria. The upshot is that much of contemporary biology – especially areas dealing with fundamental questions about the nature and origin(s) of life – is founded upon an unrepresentative subsample of a single example of life. In this light, Chapter 6 explores possibilities for reframing biological theory from a microbial standpoint. While a more microbial centered biology will not solve the infamous $N = 1$ problem, basing a theoretical framework for life on the most representative form of Earth life holds forth the promise of achieving a better understanding of familiar life. And it just might pave the way to a better foundation for exploring the possibilities of life elsewhere. Astrobiologists concur that most life in the universe is microbial and moreover probably bacteria-like.[3]

A pressing problem remains: How can scientists conduct a search for unfamiliar life without a definition or theory of life to guide it? Chapters 7, 8 and 9 explore three strategies for overcoming this problem: (1) artificially creating novel forms of life right here on Earth (Chapter 7), (2) searching for extraterrestrial microbial life by identifying potentially biological anomalies using tentative (versus defining) criteria for life (Chapter 8), and (3) searching for a "shadow biosphere" (microbial life on Earth descended from an alternative biogenesis) (Chapter 9). As will become apparent, the first (artificial life) strategy is too closely based on current Earth-centric concepts of life to tell us much about the possibilities for truly

---

[3] Even Dirk Schulze-Makuch and William Bains (2017), who provocatively argue that complex multicellular life is far more common in the universe than most scientists believe, would concede this point since they do not disagree that the first living things to arise from nonliving chemicals are unicellular and moreover that their unicellular descendants continue to flourish as life becomes increasingly more complex.

different forms of life. What is needed to make progress on the $N = 1$ problem are samples of *natural* life descended from an alternative biogenesis. This strategy is the topic of Chapters 8 and 9. Chapter 8 advocates searching for extraterrestrial life using tentative (versus defining) criteria based on familiar life. The function of tentative criteria is not (like a definition) to decide the question of life, but instead to identify promising candidates for further, more focused, scientific investigations. Such candidates will manifest as anomalous, that is, as resembling familiar life in provocative ways and yet also differing from it in unanticipated ways. In this context, Chapter 9 explores the intriguing possibility that we may not need to look beyond Earth to discover novel life; our home planet may be host to as yet unrecognized forms of life descended from an alternative biogenesis.

1

The Enduring Legacy of Aristotle:
The Battle over Life as Self-Organization or
(Genetic-Based) Reproduction

1.1 Overview

There are universal theories in physics and chemistry but no universal theories in biology. The failure of biologists to come up with such a theory is not due to a lack of effort. Philosophers and scientists have struggled to formulate universal principles of life since at least the time of Newton. This chapter traces the history of these efforts back to their roots in the work of the ancient Greek philosopher Aristotle. Aristotle's influence can be seen today in the view, which dominates contemporary biological thought about the nature and origin(s) of life, that the following abstract functional characteristics are basic to life: (1) the capacity to self-organize and maintain self-organization for an extended period of time against both external and internal perturbations and (2) the capacity to reproduce and (in light of Darwin's theory of evolution) transmit to progeny adaptive characteristics. For the sake of simplicity, I refer to the former as "O" and to the latter as "R" throughout this chapter. As Section 1.2 discusses, the conceptual parallels between O and R and Aristotle's ideas about life are remarkably close. He identified "nutrition" and "reproduction" as the basic functions of life and debated (as do so many contemporary researchers) which is more basic. Aristotle also bequeathed to biology the thorny problem of teleology – the notion that the allegedly basic functions of life (in their contemporary guise, metabolism and genetic-based reproduction) require a strange (to the modern scientific mind) form of causation that is intrinsically directed at achieving a future goal. As Aristotle argued, living things are not just fed, they feed *themselves*, and they are not just copied, they reproduce *themselves*. Characteristic O reflects this view in explicitly referring to the idea of self-organization. Similarly, characteristic R implicitly assumes that organisms contain an internal principle for generating organisms resembling themselves; external processes do not (like a 3D printer) duplicate them.

Sections 1.3 and 1.4 trace how Aristotle's seminal ideas about life evolved into characteristics O and R. As will become clear, the idea that the causal processes responsible for the distinctive functions of life are intrinsically goal directed poses special challenges to scientific reasoning. Teleological causation is not easily accommodated within the framework of classical physics, which holds that causes precede their effects without anticipating them. With the advent of Darwin's theory of evolution by natural selection, many biologists became convinced that the *prima facie* teleological properties of life could be explained in terms of undirected cause and effect relations after all. As we shall see, this claim is too sweeping. But even supposing that teleological causation has been exorcised from biology, the conceptual parallels between Aristotle's ideas about life and characteristics O and R remain uncannily close. From a philosophical perspective, the parallels between Aristotle's ideas and characteristics O and R are intriguing because, as the history of science reveals, in other domains of scientific inquiry (e.g., chemistry and physics) the most rapid advances occurred after the abandonment of Aristotelian concepts and principles.

The second half of the chapter (Section 1.5) explores the Aristotelian character of contemporary scientific thought about the nature and origin(s) of life. Theories of life are commonly articulated in the form of definitions. Following in the footsteps of Aristotle, definitions of life invariably privilege one of the characteristics O or R over the other (Section 1.5.1). Those privileging characteristic O (metabolic definitions) struggle with explicating a philosophically coherent and scientifically fruitful concept of self-organization, frequently falling back upon concepts (e.g., autopoiesis) that sound suspiciously teleological.

Theories (often called "models") of the origins of life divide along the same lines as definitions of life with regard to characteristics O and R (Section 1.5.2). Tacit appeals to causal processes that are self-generating and goal directed are routine, for example, the "spontaneous assembly" of chemically improbable, primordial biomolecules, such as small proteins (peptides) or small RNA molecules, from more basic molecular components, and the "emergence" of protoorganisms from complex, autocatalytic, chemical reaction networks. In other words, not only do O and R closely parallel Aristotle's account of life in terms of nutrition and reproduction, they are also difficult to explain in terms of ordinary, undirected causal processes. Some origins of life researchers attempt to circumvent this difficulty by arguing that such events are merely extraordinarily improbable, as opposed to being the products of a weird form of causation. As we shall see, however, this strategy is just as unscientific as one that embraces teleological causation.

The purpose of this chapter is to motivate the thesis that contemporary thought about life may be being held hostage to neo-Aristotelian ideas.[1] For this reason it covers a lot of ground (historical, philosophical, and scientific), much of which will be revisited in greater depth in subsequent chapters. Let me hasten to add that I am not claiming that a neo-Aristotelian framework for theorizing about life is mistaken. Although it has thus far been unsuccessful in providing a general theory of familiar Earth life, let alone life considered generally, and is fraught with conceptual problems, it might still turn out to be the best approach. I am arguing that we should stop *reflexively* viewing life through Aristotelian lenses. Before giving up on the prospects for universal biology, as so many pluralists recommend (see Chapter 6), we need to explore the possibility that the ostensible lack of unity among biological phenomena is the result of tacit commitment to a defective, neo-Aristotelian, theoretical framework for reasoning about life. For as Chapter 4 discusses, an unpropitious set of basic theoretical concepts can frustrate the search for unity among a diverse body of natural phenomena *even when it exists*; not every way of carving up a domain of natural phenomena into fundamental categories is capable of yielding empirically powerful, unifying generalizations.

1.2 Aristotle on the Nature of Life: Nutrition Versus Reproduction

The history of thought about the nature of life is rich and complex. I cannot do it justice here. My purpose is to trace the development of a few, highly influential, core assumptions about life in a selective manner.[2] Chief among them is the idea that the distinctive features of life are the product of a special form of causation not found in inanimate material systems. This supposition is important because, as we shall see, it underlies functional characteristics O and R and, most importantly, has reappeared in contemporary scientific debates over the nature and (especially) the origins of life. Our historical journey begins in ancient Greece with the writings of the philosopher-scientist Aristotle. As will become apparent, in almost every domain of natural science except speculation about the nature and origins of life, Aristotle's views have been rejected.

Aristotle (384–323 BC) is credited with being the first biologist because he emphasized the importance of basing theoretical conjectures about life upon empirical investigations.[3] Although limited to what could be seen with unaided human vision (primarily large, complex organisms, viz., plants and animals),

---

[1] A recent resurgence of neo-Aristotelian ideas about life among philosophers (e.g., Bedau 2010; Groff and Greco 2013) illustrates the continued influence of Aristotle's ideas.

[2] See Coleman (1977), Grene and Depew (2004), and Sapp (2003) for detailed and authoritative discussions of the historical development of biology.

[3] See Guthrie (1990, Ch. II) for more information on the life of Aristotle.

Aristotle engaged in detailed and systematic investigations of a wide variety of organisms. He observed animals in their natural habitats, studied their development, experimented on them, dissected them and even vivisected them. For these reasons, he is widely and appropriately known as the father of biology. During a period of self-imposed exile on the island of Lesbos, Aristotle spent several years conducting careful studies of marine animals in tide pools. Here he met and collaborated with Theophrastus, a native of the island, who was interested in plants. Aristotle invited Theophrastus to join him in Athens at the Lyceum, where they continued their collaboration, with Theophrastus focusing on plants and Aristotle on animals. Many of Aristotle's works on biology (*History of Animals*, *Parts of Animals*, *Generation of Animals*, as well as shorter essays on animal locomotion, respiration, etc., and the highly theoretical *De Anima*) survive to this day, providing a rich source of information about his views about life; see Barnes's *The Complete Works of Aristotle* (1984).

In his most theoretical work in biology, *De Anima* (trans. Barnes 1984 and also Shields 2013; reprinted in Bedau and Cleland 2010), Aristotle discusses the nature of life, distinguishing living things from nonliving things in terms of functional capacities such as nutrition, reproduction, sensation or perception, locomotion, and thinking (*De Anima* Bks I and II). He identifies nutrition and reproduction as more basic than the others. For Aristotle, nutrition is not nutrients (food) but rather the *internal* capacity or "power" [*dynamis*, in ancient Greek] of an organism to acquire (absorb), process (digest), and use nutrients for biological ends such as development, growth, maintenance, and repair; tellingly, he uses the terms "self-nutrition" and "nutrition" interchangeably in his writings. In *The Parts of Animals* (Bk I 639^{b15}–639^{b32}), Aristotle draws an analogy between a builder constructing a house and the development of an organism, arguing that the causal processes involved can only be understood in terms of their end products, respectively, a finished house and a mature organism; the activities of both the builder and the organism are prearranged in such a way as to bring these ends about. He also observes that there is a significant difference between "productions of nature" and the constructions of human beings. Buildings do not construct themselves. The causation involved is external to the material used to construct them, residing in the intentional (according to a plan) causal activity of human beings. In contrast, the causal processes responsible for constructing and maintaining an organism are internal; organisms build and regenerate *themselves*. It is as if the end product (a mature, thriving organism) is already present in ghostly form, actively organizing and guiding the complex causal processes required to produce and maintain it. Such processes are difficult to make sense of in terms of the mechanistic conception of causation familiar from classical physics. Causes of the latter sort (e.g., wind) do not anticipate or somehow represent (analogous to the plan of a human builder) their effects (falling leaves) in advance of their occurrence.

To account for the distinctive functional characteristics of life, Aristotle postulated a special form of natural causation (*De Anima* Bk I). Living things are *self-causing* in the sense that they have *internal* regulative principles or "souls" (*De Anima* Bk 1.4).[4] The acquisition of a soul by an inanimate object creates an "organized being" having the capacity to "move" its parts ("organs") in such a way as to engage in a certain type of functional activity (nutrition, reproduction, locomotion, etc.) as a whole (*De Anima* Bk II 412^{a28}). Nonliving phenomena are not self-causing. Lacking the restorative nutritive power, a mountain cannot sustain its massive form against the processes of erosion; subject to the action of wind and water, it becomes smaller and smaller, eventually disappearing. Analogously, lacking the nutritive power, the maintenance of buildings requires the highly organized, external activity of human beings. Living things are different. They have causally efficacious, internal principles of organization ("soul powers").

Whereas modern science accepts only one form of causation, Aristotle distinguished four: efficient, material, formal, and final (*Physics* Bk II 1–8). The only one with an analog in modern science (and only imperfect or approximate at that) is efficient causation, which proceeds linearly in time from cause to effect. Final causes, in contrast, act in the opposite direction, shaping and directing occurrent processes towards future goals that are somehow anticipated in advance.[5] To the extent that the activities generated by a soul power are directed at achieving a certain end, it is both an efficient and a final cause. As an illustration, consider the development of a fertilized egg into a mature organism. That a particular zygote will develop into a zebra – as opposed to, say, a frog – is predetermined. The causal processes involved unfold as one stage (efficiently) causes the next, but always in such a way as to (finally) produce an adult zebra. Importantly, there is no *external* guiding hand. It is as if the zygote contains within itself a self-following recipe for making a zebra.[6]

The most important of the life functions (nutrition and reproduction) are unambiguously teleological. The internal activities that realize them are directed at

---

[4] Aristotle's concept of "soul" (or *psuchê*, in ancient Greek) is very different from the modern concept. A soul is not a divine entity that survives death, nor does it presuppose consciousness or intelligence. It is a principle (or, as it is translated, "breath") of life, i.e., a cause or power that confers a life function on a material thing (Claus 1981). According to Aristotle, different life functions are associated with different souls.

[5] Some scholars of Aristotle (e.g., Wieland 1975) contend that his "four causes" are not genuine causes but rather something closely related, viz., forms of explanation. For our purposes, however, we do not need to settle this issue about Aristotle. The important point is that neo-Aristotelian accounts of how life differs from nonlife interpret them as causes, distinguishing final (goal-directed) causation from efficient causation. Much of the modern literature on emergence, for instance, is dedicated to fleshing out a coherent concept of how a cause can be goal directed.

[6] Viewed in light of what we have learned about how hereditary information is stored on DNA and translated into proteins for use in constructing organismal bodies, Aristotle's account sounds remarkably contemporary!

bringing about particular ends, for example, a mature, thriving organism. But, while self-generated activities and goal-directed activities often coincide, they need not. Indeed, for Aristotle, these functions may come apart in the case of locomotion (Furley 1994).[7] The locomotion of an antelope, for instance, could be just for the sake of moving; it need not have a definite purpose, such as fleeing from a lion or seeking food. Nevertheless, the motion of the antelope is self-caused in the sense that its causal source lies *within* the antelope, arising from the internal coordination (or, in Aristotelian terminology, "movement") of nerves, muscles, and sinews. Thus, just as being goal directed is no guarantee of being self-caused (e.g., the construction of a building is goal directed but not self-caused) so it seems that being self-caused is not, for Aristotle, a guarantee of being goal directed. Aristotle's account aside, however, there is a weaker sense in which the internal activities of a running antelope are surely goal directed: Their "purpose" is to generate *spontaneous* motion in the antelope. In contrast, the activities that inanimate objects participate in are neither goal directed nor self-generated. A stone is incapable of spontaneous motion; only an external cause, for example, rapidly flowing water or the motion of a foot, can move it. We will return to the distinction between being self-caused and being goal directed shortly. As will become apparent, some of the contemporary debates over the nature and origins of life implicitly appeal to it.

Impressed by the hierarchical functional organization of the parts of animals and plants, Aristotle classified organisms in terms of combinations of life functions. All living things (plants and animals) share the most basic soul powers, (self-) nutrition and reproduction. In this context, it is important to distinguish Aristotle's concept of reproduction from his notion of *regeneration*, which he would classify under nutrition because it involves the capacity of an existing organism to sustain and repair itself. Reproduction, in contrast, is concerned with the internally orchestrated "production of another like itself" for the purpose of preserving the species of which it is a part (*De Anima* Bk 2.4 415^{a2}; trans. Barnes 1984, 661). Some living things have additional life functions and can be classified taxonomically according to which ones they have and which ones they lack. As an illustration, animals have sensation and humans have, in addition to the soul powers of animals, the capacity for thought (*De Anima* Bk 2.3 414^{a30}–414^{b19}; trans. Barnes 1984, 659). On Aristotle's account, the other life functions presuppose reproduction and nutrition, for example, sensation cannot occur without nutrition.

---

[7] For Aristotle, self-causation amounts to internally generated "movement" of the parts of an organism, which gives rise to vital functions such as nutrition and locomotion (*De Anima* Bk 1.4). It is thus important to distinguish Aristotle's concept of locomotion (change in place of the whole organism) from his concept of movement (*qua* self-causation).

Scholars disagree about whether Aristotle believed that one of the basic life functions is more fundamental than the other. In some passages he suggests that nutrition is more fundamental: "It follows that first of all we must treat of nutrition and reproduction, for the nutritive soul is found along with all the others and is the most primitive and widely distributed power of soul, being indeed that one in virtue of which all are said to have life" (*De Anima* Bk 2.4 415^{a22}–415^{a26}; Barnes 1984, 661). Yet, as Gareth Matthews (1992) observes, there are also passages in which Aristotle suggests that reproduction is the more fundamental life function because "the production of another like itself ... is the goal towards which all things strive, that for the sake of which they do whatsoever their nature renders possible" (*De Anima* Bk 2.4 415^{a28}–415^{a32}; Barnes 1984, 661). Matthews concludes that Aristotle is best interpreted as holding that "generation" (reproduction) is more essential to life than nutrition, and moreover that he literally "defines" life in terms of generation. Aristotle's writings support this view: He speaks of "defining" the different types of soul (which, as mentioned above, distinguish living things from nonliving things) in a manner analogous to the geometer's paradigmatic definitions of 'square' and 'triangle' (*De Anima* Bk 2.3 414^{b20}–414^{b33}; trans. Barnes 1984, 659). Nonetheless some scholars of Aristotle, for example, Shields (2002), contend that he was not concerned with defining life.[8]

For our purposes, however, the important point is that there are close conceptual parallels between Aristotle's functional characterization of life, in terms of self-nutrition and self-reproduction, and contemporary characterizations of life, in terms of O and R. Aristotle explicated self-nutrition metabolically, in terms of the internal capacity of organisms to absorb and process "food" found in the environment for the purpose of bringing about and "... maintain[ing] the [organized] being of that which is fed" (*De Anima* Bk 2.4 416^{b1}–416^{b17}; trans. Barnes 1984, 663). He characterized reproduction as species preserving to the extent that it involves an organism generating another organism that closely resembles it. Indeed, as Ernst Mayr (1988, 55–60) observes, Aristotle's discussions of reproduction, development, and adaptation in living organisms sometimes sound uncannily modern. What does not sound modern is the idea of goal-directed self-causation as the ultimate explanation for the difference between living and inanimate material things. Yet as adumbrated above, and discussed in greater detail below, it is difficult to articulate characteristics O and R without at least tacitly invoking it. The question is can this unsettling form of causation be expunged from O and R?

---

[8] I find this view implausible. Aristotle founded the discipline of logic, and wrote extensively about the logical character of definition, which suggests that he would not use the Greek word for definition sloppily. Still, I am not a scholar of Aristotle, and there is some disagreement over this issue. (Perhaps, and this is pure speculation, "nutrition" captures self-causation and "reproduction" the goal-directed aspects of life; since, as discussed earlier, the two concepts are not the same.)

1.3 Classical Mechanism about Life: From Optimism to Quiet Desperation

Aristotelian ideas explicitly dominated scholarly thought about life until the writings of René Descartes (1596–1650).[9] Influenced by the rapidly developing mechanistic account of nature, which culminated in the seventeenth century with Isaac Newton's laws of motion. Descartes argued, in "Treatise of Man" (trans. Hall 2003; reprinted in Bedau and Cleland 2010), that living things are automata, albeit much more complex than the artificial devices powered by water, air, and wound springs popular during his day; the bones, muscles, and organs of animals were likened to cogs, pistons, and cams. The only analog to Aristotle's four causes in Newtonian physics is force (efficient causation); material causation is relegated to passive matter, and formal and final causation are completely absent.

But in the ensuing years, as scientists learned more about plants and animals, it became clear that organismal bodies differ in important ways from mechanical devices. Vitalism, the view that life involves a special kind of force, energy, or substance, was a popular response to these difficulties.[10] The most influential versions of vitalism (e.g., that of Jöns Jakob Berzelius) attempted to accommodate life within Newton's highly successful theoretical framework by postulating a special life force for maintaining the dynamical functional organization of living things. Such an approach seemed eminently reasonable. Newton's three laws of motion applied to forces considered generally. They left open the identity of particular forces, including the possibility that new forces would be discovered.

Worries about whether life could be accommodated within the theoretical framework of classical physics culminated in the late eighteenth century with the writings of Immanuel Kant (1724–1804), who famously concluded that there would never be a Newton of biology (*Dialectic of Teleological Judgment* §75; trans. Guyer and Matthews 2001; reprinted in Bedau and Cleland 2010). Kant argued that life could only be understood on the assumption that it is produced according to a design; organisms are intrinsically purposive or goal-directed entities. The source of this design, however, is not an intelligent designer but rather nature.[11] How is it possible for design to exist in nature without an intelligent designer? According to Kant, the very concept of something being a natural end requires conceiving of it as self-causing, that is, as "both cause and effect of itself" (§64, 371). Kant also argued that life is unique in being naturally teleological. In short, for Kant, organisms differ from inanimate objects in having internal, purposive, regulative principles that cannot be explained within a

---

[9] See Grene and Depew (2004) for more on the evolution of thought about the teleological properties of life from Aristotle to contemporary times.

[10] For an authoritative history of vitalism, see Coleman (1977).

[11] Kant was a well-known astronomer as well as a philosopher. He is credited with developing the nebular hypothesis, which explains how planets and stars form from gaseous clouds (nebulae).

Newtonian framework of forces (vital or mechanical); echoing Aristotle, he dubbed living things "organized beings." Thus, Kant (an avid Newtonian, by the way) brought our understanding of life full circle back to a more refined version of Aristotle's view.

Kant's argument was subtle, too subtle for some vitalists, who took it as vindicating their views (Mayr 1997, Ch. 1), and vitalist ideas about life remained popular throughout the nineteenth century and into the early twentieth century; the last eminent biologist with vitalist leanings, Hans Driesch, died in 1941. Other philosophers, for example, J. S. Mill (1843), and scientists, for example, C. Lloyd Morgan (1912), however, recognized that Kant was attacking the foundations of the Newtonian picture of the causal structure of reality.[12] They proposed an alternative to vitalism that has come to be known as emergentism. According to emergentists, there is no special vital substance. Everything (living and nonliving) is made up of the same physical stuff. At certain levels of complexity, however, novel properties and regularities (patterns) arise that cannot be analyzed in terms of the laws of fundamental physics.

As with vitalism, there are different versions of emergentism. The so-called British emergentists were the first to attempt a comprehensive philosophical theory of emergence, detailing its logical structure and how it challenges the picture of causation inherited from classical physics.[13] A brief overview of British emergentism is therefore instructive. In classical physics, forces act linearly on physical objects and are subject to the principles of vector composition and decomposition. All forces, however complex, are fully decomposable into basic forces (e.g., gravitational and electrical) operating upon pairs of particles of the appropriate kinds (in the cases at hand, mass and charge, respectively). It is here that the British emergentists break with classical physics. The basic particles that make up material objects are collectively organized into increasingly complex structures. Sufficiently complex structures exhibit causal powers that cannot be analyzed as net forces and reduced to those of fundamental physics; the whole is literally greater than the sum of the parts.

According to British emergentism, there is a hierarchy of these causally efficacious, physical structures, beginning, in ascending order, with strictly physical kinds, chemical kinds, biological kinds, and psychological kinds. Each level is

---

[12] As an avid Newtonian, Kant worried about this and waffled on whether teleological causation represents a fundamental feature of life or whether it represents a mere limitation of the human intellect; he speculated that an alien intelligence might be able to understand life within the framework of classical physics but that humans could not cognize life nonteleologically. It is beyond the scope of this book to pursue the subtleties of Kant's account; for more on Kant's views about the teleological character of life, see McLaughlin (1990).

[13] British emergentism reached its high point in the early twentieth century with the work of philosopher C. D. Broad (1925). For an informative discussion of the history of emergentism, in general, and British emergentism, in particular, see McLaughlin (1992).

characterized by distinctive, causally efficacious, properties that depend upon but cannot be reduced to those occurring at a lower level. As McLaughlin (1992) points out, these (level dependent) novel causal properties amount to "configurational forces." They can only be exercised by configurations as wholes. Some configurational forces act downwardly on the components of the system, shaping and directing lower-level causal activities. Thus, according to emergentists, complex physical wholes at the biological level (a living antelope) can behave in novel ways (e.g., locomote) in virtue of orchestrating causal interactions among their parts (nerves, muscles, sinews, and bones) that would not otherwise occur. The internally coordinated operation of these lower-level parts critically depends on the organizational integrity of the system as a whole; put starkly, dead antelope, which have lost the pertinent, finely tuned, internal organization, cannot run. The resemblance between configurational forces and Aristotle's internal regulative principles is striking.

The basic difference between vitalism and emergentism is that the former (in keeping with the theoretical framework of classical physics) seeks to explicate life in terms of novel forces generated by objects bearing special, life conferring, intrinsic properties in a manner analogous to the way in which gravitational force is generated by objects having mass and electrical force is generated by charged objects. Indeed, the association between forces and intrinsic properties of material objects (viewed as composites of basic physical particles) persists to this day. In quantum mechanics, each of the four fundamental forces is carried by a special fundamental particle, that is, a particle having a unique intrinsic causal property: Gluons carry the strong nuclear force, certain bosons the weak nuclear force, photons the force of electromagnetism, and (allegedly) gravitons the force of gravitation.[14] In contrast to vitalism, emergentism holds that living things are constituted by the same kinds of fundamental particles (substance) as inanimate objects. What distinguishes living things from inanimate matter is the manner in which these particles, and the higher-level physical objects (protons, atoms, molecules, etc.) built from them, are collectively organized. When a material system attains the requisite level of organizational complexity it acquires novel functional capacities (e.g., metabolism, locomotion, sensation, and reproduction) that cannot be explicated in terms of lower level causal interactions among more basic physical parts.

Vitalism and emergentism are sometimes (e.g., Mayr 1997, Ch. 1) grouped with theological accounts as involving "supernatural forces." This view rests upon the assumption that the only legitimate form of natural causation is ultimately mechanistic or, even more minimalistically, Humean (consisting in nothing more than

---

[14] One of the central problems of quantum theory is the lack of empirical evidence for gravitons.

the regular succession of events[15]). The question of whether all causation is Humean is surely an empirical matter, as opposed to an a priori (logical) truth. As a case in point, quantum phenomena, such as quantum nonlocality, have been held up as challenging the traditional linear concept of causation.[16] For our purposes, however, the important point is that the versions of vitalism and emergentism just discussed, which are among the most fully developed accounts, represent attempts to expand the theoretical framework inherited from classical physics in wholly naturalistic ways to accommodate long standing, seemingly intractable, puzzles about life. The real problem with both vitalism and emergentism is the extremity of the revisions required to flesh them out, namely, the introduction of primitive vital forces and their carriers (vital properties or particles) or alternatively novel forms of causation (primitive configurational forces).

1.4 Darwin to the Rescue?

With the advent of Darwin's theory of evolution by natural selection in the mid-nineteenth century (Darwin 1859), many biologists and philosophers believed that they had found a way of explaining the ostensibly purposive (teleological or goal-directed) characteristics of life in terms of causal mechanisms familiar from classical physics; as Darwin's friend Thomas Huxley gloated, "... Teleology, as commonly understood, ha[s] received its deathblow at Darwin's hands" (Huxley 1864, 82). To revisit a famous example of Darwin's, finches on the Galápagos Islands have beaks of widely differing sizes and shapes that are highly adapted to their food sources. Those eating seeds off the ground, for instance, have shorter, stouter beaks, whereas those feeding on insects have sharper, more slender beaks. It is as if their beaks were deliberately designed to enhance their survival by

---

[15] The modern scientific concept of causation as *de facto* regularity was developed by the philosopher David Hume (1888) in the eighteenth century. Hume provided powerful, most importantly, empirically based, arguments against the widespread view that causation involves a mysterious, imperceptible, physical connection between causes and their effects. He argued that the relationship of cause to effect is just a matter of invariable succession (time-ordered correlation) – of events of one kind (causes, such as kicking a stone) always being followed by events of another kind (effects, such as the motion of a kicked stone). Hume's analysis of causation continues to dominate thought about causation in natural science (albeit mostly tacitly) and much of philosophy as well.

[16] Quantum nonlocality seems to involve (non-Humean) causes that do not linearly precede their effects but rather occur simultaneously with them, even when vast distances are involved; it is as if the cause "acts" at a distance rather than first being transmitted through an interval of space lying between cause and effect. While quantum nonlocality does not require internal regulative causal principles of the sort being entertained by vitalists and (biologically oriented) emergentists, it does (like them) violate the traditional mechanistic and Humean notion of causation as (ultimately) consisting in linear sequences of causes *followed by* their effects. Indeed, the disquiet of many theoretical physicists and philosophers of physics over quantum phenomena, as underscored by efforts to shore up the "hidden variable" interpretation of quantum mechanics, underscores just how influential the traditional Humean account of causation remains. Indeed, some philosophers and physicists argue that the success of quantum mechanics shows that there is no causation at the quantum level. In any case, however, it is important to keep in mind that quantum nonlocality and teleology provide very different sorts of challenges to the Humean notion of causation.

making it easier to exploit the food sources available to them. But how could nature, left to its own devices, manage something like this?

Aristotle solved this problem by introducing a natural form of goal-directed self-causation: "... it is both by nature and for an end that the swallow makes its nest and the spider its web, and that plants grow leaves for the sake of the fruit and send their roots down (not up) for the sake of nourishment ..." (*Physics* Bk II 199^{a25}; trans. Barnes 1984, 340). For Aristotle, goal-directed causation is hierarchical. The parts (organs) of organisms have ends (the survival and maintenance of the individual) and organisms, in turn, are parts of eternal species, and hence have higher order ends, namely, the survival and maintenance of the species (Vella 2008, Ch. 3). Many of Darwin's scientific predecessors, eschewing a form of natural causation incompatible with classical physics, attributed the remarkable adaptedness of organisms to their environments as powerful evidence for the existence of God; God deliberately designed them so that they could survive and reproduce in the environments in which he placed them. Darwin's theory of evolution by natural selection advanced a naturalistic explanation for the adaptations of organisms that, unlike neo-Aristotelian vitalism or emergentisim, did not presuppose teleological causation.

The core ingredients in Darwin's theory of evolution are a physical source of heritable variation and a natural mechanism for biological change, viz., natural selection. Small heritable variations are always present in a population, providing a potential source of new adaptations. If the environment changes in such a way as to render a hereditable characteristic advantageous to the survival and reproduction of an organism, the characteristic will *tend* to become more common in the population; alternatively if it is disadvantageous, it will *tend* to become less common.[17] In other words, the natural environment exerts a selective pressure for advantageous traits and against disadvantageous traits in a manner analogous to the way in which human breeders artificially select for desired traits in plants and animals. It is thus likely that organisms having advantageous traits will eventually come to dominate a population, leading to new varieties and sometimes even new species. The important point, for our purposes, is that there is nothing special about the causation involved. It is admittedly very complex, involving heritable modifications to individual organisms and many interactions of various sorts between organisms and their environments over long periods of time. As the great twentieth

---

[17] These "tendencies" concern what is statistically expected to happen given a particular *type* of population and a certain *kind* of environment, not what *actually* happens to an individual organism or particular population. An organism with an advantageous trait may fail to survive and reproduce because it has bad luck, suffers an injury, or has some other trait that is very disadvantageous. A pandemic may destroy a whole population of organisms. Darwinian accounts of biological change are thus idealized to the extent that they abstract away from such factors – factors that influence what actually happens to individual organisms and particular populations but are nonetheless extraneous to the evolutionary outcomes (fitness differences) of interest.

century biologist Ernst Mayr (1992) counseled: "Natural selection deals with the properties of individuals of a given generation; it simply does not have any long range goal, even though this may seem so when one looks backward over a long series of generations" (p. 133).

Many of Darwin's contemporaries were nonetheless convinced that biological evolution is inherently teleological. Even Darwin himself sometimes seems to endorse this view (Grene and Depew 2004, Ch. 7). The challenge for early Darwinians was explaining how an undirected, externally (environmentally) mediated, process of natural selection operating on small hereditary variations (that are random with respect to their adaptive potential) among individual organisms could explain evolutionary trends such as those exhibited in the fossil record. As a result, vitalism and emergentism continued to attract followers into the early years of the twentieth century. A related view, orthogenesis, emerged in the late nineteenth century, and also attracted followers into the early twentieth century. Orthogenesis holds that there are trends in evolution that have no adaptive significance (and hence are difficult to explain in terms of natural selection). It explains these trends by postulating an intrinsic drive (Shanahan 2004). By the middle of the twentieth century, however, teleology was being rapidly expunged from biology as a result of two critical scientific advances: The development of the Modern Evolutionary Synthesis in the 1930s–1940s and the invention of the programmable computer in the 1940s.

The Modern Evolutionary Synthesis unified ideas from a number of biological specialties, including Mendelian genetics, population genetics, embryology, ecology, and paleontology, into a comprehensive, modern evolutionary theory.[18] Heritable variation, upon which natural selection operates, was fleshed out genetically in terms of causally unproblematic notions, viz., mutations that are random (versus directed) with respect to adaptation, recombination, and genetic drift, and the physical basis of heredity, which had evaded Darwin and his contemporaries but was pursued chemically, eventually culminating in the discovery of DNA (Sapp 2003, Ch. 12). Advances in ecology and paleontology, coupled with studies in population genetics, provided scientifically compelling accounts of how natural selection operating on small genetic variations among organisms in natural environments could produce new adaptations and eventually new species over geologically plausible time scales (Grene and Depew 2004, Ch. 9).

With the advent of the programmable computer it became easier to understand how genetic information stored chemically in chromosomes could direct the development of inherited phenotypic (morphological, metabolic, and even behavioral) characteristics of individual organisms. The influence of computer science on

---

[18] See Huxley (1942) for the now classic statement of the theory.

the development of molecular genetics was profound (Jacob 1973). An organism's genome was likened to a computer program, shaped by natural selection and inherited from its parents. This "genetic program" literally "encodes" all the information required for an organism to construct itself and to continuously regenerate (maintain and repair) itself over the course of its life. The development of the computer metaphor, in the context of rapid advances in molecular biology, left little doubt in the minds of biologists, paleontologists, and philosophers of biology that the ostensibly goal-directed characteristics of life that so impressed Aristotle, and perplexed his eighteenth and nineteenth century successors, could be explicated without presupposing a novel form of causation.

Fleshed out by the Modern Evolutionary Synthesis and the computer metaphor, Darwin's theory does an impressive job of explaining the unity and diversity of known life on Earth. But as discussed below, the fact that it can successfully explain otherwise puzzling relationships among organisms on Earth does not establish that the ability to evolve in this manner is universal let alone essential to life. In addition, Darwin's theory says nothing about how life arises from abiotic chemicals under natural conditions. For, as the title of his *magnum opus* (*On the Origin of Species*) underscores, Darwin's theory *presupposes* an advanced form of life already divided into biological taxa (species) and capable of adaptive evolution.[19] Put briefly, although his contributions to our understanding of life *on Earth* are truly monumental, it is not clear that Darwin is the long-sought Newton of biology.

1.5 Here We Go Again: Aristotelian Roots of Contemporary Accounts of the Nature and Origin(s) of Life

The influence of Aristotle's ideas on contemporary biological thought is most apparent in attempts to answer fundamental questions about life, especially: What is life? How does life arise from nonliving chemicals under natural conditions? Contemporary theories of the nature of life, which are usually articulated as definitions,[20] are almost always founded upon characteristic O or R, with one being advanced as more fundamental than the other. Similarly, theories of the origins of life tend to divide along the same lines as definitions of life with regard to these characteristics. As discussed below, however, it is not at all clear that either O or R provides a scientifically fruitful, theoretical foundation for exploring

---

[19] Creationists who criticize Darwinian evolution because it cannot explain the origins of life either fail to understand this or deliberately suppress it.

[20] As discussed in Chapter 4, definitions and scientific theories are different sorts of things; good definitions supply necessary and sufficient conditions whereas good scientific theories are much more open ended. For our purposes here, however, this issue is irrelevant.

deep-seated questions about life. Moreover, even supposing that they do, the perennial problem of goal-directed self-causation remains, for both O and R seem to presuppose it.

1.5.1 Metabolism-Based Versus Evolution-Based Definitions of the Nature of Life

The focus of this section is on the Aristotelian roots of the most influential definitions of life. Most contemporary definitions of life are either metabolism based (characteristic O) or evolution based (characteristic R). Some combine both characteristics in a composite definition, for example, Dyson (1999). Following Aristotle, however, there is a tendency to treat one of these characteristics as more fundamental to life than the other; see the following chapter (Section 2.2) for more detail on these definitions.

Metabolism-based definitions characterize life as a self-organizing system having the capacity to sustain itself against degrading processes for an extended period of time by extracting material, energy or information from its environment. The most popular metabolic definitions can be classified as chemical-metabolic, thermodynamic, or autopoietic. These categories correspond to different levels of abstraction, with chemical-metabolic definitions being the most closely tailored to the metabolic processes of familiar Earth life and autopoietic definitions being the furthest removed. On a chemically based metabolic definition, life is a chemical reaction system that sustains itself by extracting and transforming chemical energy from its environment (e.g., Gánti 2003; Kauffman 1993). Most versions do not require that life uses the same biomolecules and metabolic pathways as familiar life. Some restrict life to organic chemicals. Other versions are open to the possibility of life based on inorganic molecules, for instance, silicon instead of carbon. All chemical-metabolic definitions, however, treat life as a fundamentally chemical phenomenon.

Thermodynamic definitions of life are more abstract. The emphasis is on the physics (versus chemical composition) of the system, namely, its capacity for maintaining bounded regions of local order by extracting energy from thermodynamic gradients found in the environment. Stuart Kauffman (1995, 2000) characterizes such systems as collectively autocatalytic, chemical reaction networks that are far from equilibrium, and argues that they cannot be understood without invoking a new law of thermodynamics. I will have more to say about this shortly.

Chemical-metabolic and thermodynamic definitions differ as to whether life is essentially a chemical or thermodynamic phenomenon. Autopoietic definitions, on the other hand, abstract away from physical as well as chemical detail.

The emphasis is on the "logic" of self-organization, that is, on how the constitutive parts of a system are structurally organized and integrated into a self-sustaining unit. The transition from a nonliving system to a living system is characterized as involving a form of "self-reference" (self-generation, self-maintenance, etc.) that cannot be reduced to relations among its components (e.g., Luisi 1993; Maturana and Varela 1973). Because the focus is on abstract organizational features of a system, autopoietic definitions are sometimes classified as "computational" (Popa 2004, 189). Any system, abstract or concrete, exhibiting the pertinent self-referential, organizational characteristics is (by definition) a living entity. Autopoietic definitions have been used to justify some very radical claims – that ecosystems, human institutions (e.g., corporations), and even Earth itself are individual living things (Lovelock and Margulis 1974; Margulis and Sagan 1995).

Evolutionary definitions of life also come in varying degrees of abstractness. Chemical-evolutionary definitions tend to be Darwinian. Most of them restrict life to nucleic acid-like organic molecules.[21] NASA's "chemical Darwinian definition," which holds that life is a "self-sustaining chemical system capable of Darwinian evolution" (Joyce 1994, xi–xii), is the most widely cited of these definitions. Some evolutionary definitions, however, treat life as a more general statistical-thermodynamic (versus chemical) phenomenon, for example, England (2013) and Goldenfeld and Woese (2011). The most abstract evolutionary definitions conceive of life as purely informational (versus physical or chemical), for example Korzeniewski (2001), Langton (1989), and Trifonov (2011). Darwinian evolution is analyzed as a purely logical process (in essence, a complex algorithm) which in principle could be realized by many different types of systems, all of which *ipso facto* qualify as living things. Evolutionary definitions of this sort resemble autopoietic definitions in abstracting away from chemical and physical detail, but as Section 2.2 discusses, Darwinian evolution is easier to capture in an algorithm than autopoiesis. Algorithmic versions of Darwinian evolution provide the main impetus for the more radical claims of artificial life (ALife) researchers, namely, that they have or soon will create living "creatures" in the information structures of computers; see Section 7.2. The most widely accepted mechanism of evolution is natural selection, but more general mechanisms – for example, Bedau's (1998) "supple adaptation," which encompasses not only natural selection but also, for example, Lamarckian selection – are sometimes invoked in evolutionary definitions of life.

---

[21] Following Schrödinger (1944), who famously anticipated the structure of DNA when he argued that hereditary information must be encoded on an "aperiodic solid" (p. 5), it is widely thought that all chemical genetic systems will consist of aperiodic polymers, and that (due to some unusual chemical constraints) molecules having a nucleic acid-like structure are the only ones capable of supporting evolution by natural selection (Benner and Hutter 2002).

The concept of metabolism underlying metabolic definitions of life is strikingly similar to Aristotle's notion of self-nutrition. Metabolic definitions emphasize the capacity of a living system to sustain *itself* as an "organized being" by extracting and processing "nutrients" (material, energetic, or even purely informational) found in its environment. The influence of Aristotle's ideas on evolutionary definitions is more tenuous. Both Darwin and Aristotle take reproduction to be essential to life. However, their account of the relation between reproduction, a process undergone by individual organisms, and similarities and differences among groups of organisms, such as species, differs. Writing two thousand years before Darwin, Aristotle believed that species are eternal. He focused on the role of reproduction in producing descendants closely resembling their progenitors. As part of an eternal species, the individual organism has an internal, goal-directed, drive to reproduce so that it can sustain the species to which it belongs. Contemporary evolution-based definitions of life, in contrast, attribute similarities and differences among groups of organisms to an undirected, externally mediated, process of differential survival and reproduction; organisms best adapted to a given environment are more likely to survive and leave descendants than those that are less well adapted.

Individual organisms are the paradigmatic living things. Evolution-based definitions of life, however, have trouble classifying them as such. Organisms do not evolve as individuals; they merely live and die. Some proponents of evolutionary definitions bite the bullet and simply deny that individual organisms qualify as living things. Others (e.g., Bedau 1998; Benner 2010) treat them as subordinate cases of life; see Section 2.2.3 for a discussion. Metabolism-based definitions, in contrast, admit the possibility of living things that are incapable of reproducing. For our purposes, the important point is that metabolism-based definitions of life construe the individual organism as the basic unit of life whereas evolution-based definitions take groups of related organisms (populations, species, or lineages) as the basic unit of life. This represents a fundamental difference in the theoretical frameworks underlying metabolic and evolutionary definitions of life – a difference presaged by the old Aristotelian controversy over whether nutrition (maintenance of the individual) or reproduction (continuity of the species) is more essential to life.

As we have seen, there is a long tradition of framing the nature of life in terms of either characteristic O or R, but not both. In truth, however, there is little contemporary scientific support for this rivalry. With the exception of viruses, whose status as living is controversial, all known life exhibits both characteristics. Moreover, in familiar life, these functions are deeply entangled at the molecular level. The biomolecules realizing characteristics O and R are different in type: Proteins supply the bulk of the structural and catalytic material required for building, maintaining, and repairing bodies, whereas a very different molecular species, nucleic acids, stores (DNA) and manages (RNA) the hereditary material required for genetic-based reproduction. Minuscule, very complex molecular machines (ribosomes), composed of both RNA and protein, which are found in large numbers in every cell, coordinate these functions. The information stored in DNA is transcribed onto RNA and subsequently translated into proteins by ribosomes. This complex, highly organized, molecular architecture provides the physical basis for metabolism, development, reproduction, and evolution. In other words, from a molecular perspective, there is little reason to suppose that characteristic R is more fundamental than O, or vice versa.

But even supposing that O and/or R were universal to known life, it does not follow that they are fundamental to life. They could be symptoms of more basic, but as yet unknown, properties of life. As an analogy, consider water. Various perceptible characteristics thought to be universal to water (e.g., being wet, transparent, and a good solvent) were traditionally used to distinguish it from other chemical substances. But this information does not reveal the underlying nature of water, namely, that it is composed primarily of molecules of H_2O. Moreover, the latter information cannot even be inferred from its triple point, a distinctive combination of pressure and temperature at which its gas, liquid, and solid phases coexist in equilibrium, which does uniquely distinguish water from all other (pure) chemical substances. The point is the modern scientific account of water required the development of a new theoretical framework for understanding chemical substance. The old Aristotelian concept of chemical substance, which dominated scientific thought until the pioneering work of Lavoisier, in the late eighteenth century, held that water is a basic (indivisible) chemical element. Lavoisier (1783) challenged this view by presenting compelling empirical evidence that water is a chemical compound, paving the way for the development of molecular theory.

Viewed in light of the above discussion, one cannot help but wonder whether characteristics O and R could be misleading us about the nature of life. That is, are characteristics that are merely symptomatic of life being elevated to the status of essential characteristics because they are thought to be universal to life? The situation is actually a bit more worrisome. For as Chapter 5 discusses, there is compelling empirical and theoretical evidence that characteristics O and R may not only be misleading but also are unreliable for discriminating living from nonliving systems. It could turn out that O and R are neither necessary nor sufficient for life.

1.5.2 Metabolism-First Versus Genes-First Theories of the Origin(s) of Life

Scientific theories of the origin(s) of life typically bifurcate along the same lines as definitions of life into genes-first models, which focus on the capacity to reproduce and undergo evolution by natural selection (characteristic R), and metabolism-first models, which focus on the capacity to self-organize and maintain self-organization for an extended period of time (characteristic O).[22] The most influential versions are the RNA World, which is a genes-first theory, and the Small Molecule (SM) World, which is a metabolism-first theory.[23] As in the previous section, our concern here is with the Aristotelian roots of these theories; they are discussed in much greater detail in Section 5.4.

Proteins are the primary catalysts used by the metabolic processes of familiar life. They consist of long chains (polymers) of amino acids. According to the SM World,[24] life originates with the development of a sufficiently complex, collectively autocatalytic, chemical reaction system involving small molecules such as amino acids, cofactors, and peptides (short chains of as few as two amino acids). But while abiotic amino acids are found in the natural environment, abiotic peptides are not. The chemical reactions required for forming peptide bonds between amino acids are thermodynamically uphill and difficult to envision occurring abiotically in a natural (versus controlled laboratory) environment. This is not the most serious challenge facing the SM World, however. The question of how a minimally autocatalytic chemical reaction network capable of increasing in complexity to the point of qualifying as proto-metabolic could arise *de novo* in an open thermodynamic system consisting of a variety of small molecules, including peptides, is not fleshed out.

An incipient autocatalytic chemical reaction network capable of developing the organizational complexity of a proto-metabolism must be able to sustain itself in the face of spontaneously degrading (thermodynamically downhill) geochemical processes.[25] The probability of a reaction network of this sort arising under natural conditions from a chemically heterogeneous and unorganized reaction system of small molecules is (given the current state of our knowledge) very small (Orgel 2008). Proponents of the SM World tend to gloss over this difficulty by appealing to an unexplained process of "spontaneous chemical self-organization." Instead of solving the problem, however, this amounts to resurrecting the old Aristotelian

---

[22] As discussed in Section 5.4.2, there is a long (and mistaken) history of thinking that one can infer an account of the origin of life from an account of the nature of life.

[23] Like the definitions of life discussed earlier, the SM World and RNA World models encompass a variety of more specific versions, differing in chemical and physical details. Chapter 5 discusses how the most popular versions of these respective theories differ among each other.

[24] Shapiro (2006) coined the term but the concept goes back to Oparin (1964) and has been developed along different lines by de Duve (1995), Dyson (1999), Kauffman (1993), Shapiro (2006), and Wächtershäuser (1990), among others.

[25] As Chapter 5 discusses, some mechanism for excluding certain molecules while admitting and concentrating others is needed. Various possibilities have been suggested, e.g., membranous compartments (e.g., micelles, aerosols, and lipid vesicles), absorbent mineral surfaces, and porous rocks, but this just pushes the problem back a step insofar as it raises the issue of how a mechanism able to achieve the sophisticated level of molecular selectivity and concentration required could arise under natural conditions; see Cleland (2013) for a more detailed discussion.

idea of goal-directed self-causation in a scientifically less objectionable form. Stuart Kauffman (e.g., 1995, 2000) is one of the few researchers who fully grasp this point. According to Kauffman, an understanding of the origin of life requires recourse to new thermodynamic principles of self-organization in complex, far from equilibrium reaction systems. When thermodynamically open reaction systems attain a certain level of complexity, they spontaneously self-organize into collectively autocatalytic reaction networks capable of sustaining themselves against internal and external perturbations and increasing their organizational complexity.

There is an intriguing parallel between Kauffman's ideas and those of the British emergentists. Indeed, Kauffman and fellow scientific travelers (e.g., Hazen 2005) routinely use the term "emergence" to characterize the processes involved in the transition from chemistry to biology. Both approaches propose specific modifications to physical theory, in the case of the British empiricists, Newtonian principles of vector analysis, and in the case of Kauffman, the foundations of statistical thermodynamics. Such a strategy has the advantage of suggesting definite ways – candidates for further scientific investigation – in which our theoretical understanding of life as a natural phenomenon may be inadequate. For our purposes, however, the important point is that both approaches are neo-Aristotelian in character. They distinguish living from inanimate material systems by invoking a natural form of goal-directed self-causation that cannot be analyzed in terms of causal mechanisms familiar to physical scientists.

In contrast to the SM World, the RNA World holds that life originated with the development of a proto-genetic system from a "prebiotic soup" of small RNA molecules and their molecular precursors.[26] The impetus for the RNA World is the Nobel Prize winning discovery that (unlike DNA) some RNA molecules are self-replicating. This discovery is held up as solving a formidable "[which came first] the chicken or egg" dilemma: The replication of nucleic acids (DNA and RNA) was previously thought to require protein enzymes and the synthesis of protein enzymes depends upon nucleic acids (Orgel 2004).

The abiotic synthesis of a small RNA molecule (RNA oligomer) under natural conditions is even more challenging, however, than the synthesis of a short peptide. An RNA molecule consists of a long chain of ribonucleotides. A ribonucleotide is made up of three molecular subunits, a phosphate unit bonded to a sugar (ribose) unit bonded to a nucleobase (purine or pyrimidine) in a precisely organized, three-dimensional pattern. The assembly of an RNA oligomer in a laboratory setting requires a long sequence of carefully staged and controlled

---

[26] The term "RNA World" was coined by Gilbert (1986) but the concept was developed earlier by Woese (1967, 179–195), Crick (1968), and Orgel (1968).

chemical reactions. Given our current understanding of geochemistry, the probability of an RNA oligomer being abiotically synthesized under natural conditions
is extremely low (e.g., Shapiro 2000, 2006). Moreover, the likelihood that such a
molecule would be self-catalytic is even lower; most RNA molecules are not self-
replicating. Indeed, the theoretical difficulties in making good chemical sense of
the synthesis of self-replicating RNA under natural conditions are so serious that
many researchers speculate that the RNA World must have been preceded by an
earlier "Pre-RNA World" based on a different self-replicating molecule (e.g.,
Joyce and Orgel 1999). Various candidates have been suggested but, as discussed
in Section 5.4, none seems much easier to synthesize than RNA.

   The central focus of the RNA World is on "evolving" a population of self-
replicating RNA molecules with increasingly efficient and versatile catalytic capacities. Most scenarios begin with a primordial pool of mutually self-catalytic RNA
oligomers. In order for the RNA World to even get off the ground, however, the
pool of molecules must have a fairly sophisticated level of chemical organization.
A pool of RNA oligomers arising under natural (versus artificial) conditions would
include a variety of other chemical species. An incipient RNA World would thus
be subject to degrading chemical reactions unless there was some mechanism for
excluding them.[27] But even supposing that it consisted exclusively of RNA
oligomers, an embryonic RNA World would be subject to cross-reactions with
nonreplicating RNA. In short, one needs just the right mix of just the right types of
RNA oligomers in a reasonably clean environment for the system to be capable of
evolving in the manner required. Proponents of the RNA World tend to downplay
this problem by appealing to the "spontaneous emergence" of self-organization in
mutually catalytic populations of small RNA oligomers. As in the case of the SM
World, however, downplaying the problem does not solve it, but instead tacitly
resurrects the concept of goal-directed self-causation in scientifically more palat-
able language.

   In addition to the problems just discussed, both the RNA World and the SM
World have difficulty explaining the origins of the deep molecular entanglement
between the metabolic and genetic systems of familiar Earth life. This difficulty is
not surprising. By positioning themselves as rivals with respect to this high level
architectural division, they in effect marginalize the issue of how these systems
became chemically integrated; their interdependence falls into the crack between
the theories. Defenders of the SM World conjecture that nucleotides and RNA
oligomers arose in a developing autocatalytic reaction network to serve some

---

[27] As in the case of the SM World, various natural mechanisms for concentrating and protecting delicate
biomolecules, in this case RNA oligomers and their building blocks, have been suggested, namely,
membranous compartments of various sorts, absorbent mineral surfaces and porous rocks. These and other
possibilities are explored Chapter 5.

nongenetic purpose and took on a genetic role much later, after the development of a primitive metabolism (e.g., Shapiro 2006). Defenders of the RNA World, on the other hand, contend that RNA molecules performed both genetic and primitive metabolic functions until some time after the appearance of the first "riboorganisms" (Benner and Ellington 1987). At some point RNA began synthesizing peptides and proteins, which being better catalysts, eventually took over the metabolic functions of RNA. Given the size and complexity of proteins and nucleic acids, however, it seems unlikely that the complex cooperative arrangement between them arose at such a late stage in the development of life. A few researchers have developed hybrid theories in an effort to deal with this problem. Copley et al. (2005), for instance, argue that amino acids became associated with the nucleobases that code for them in a small molecule reaction network much earlier, prior to the synthesis of either proteins or nucleic acids. For our purposes, however, the important point is that the ancient architectural division of life along the lines of characteristics O and R seems to be driving theorizing about the origin of life.

1.6 Concluding Thoughts

To wrap up, contemporary scientific and philosophical thought about the nature and origin(s) of life is founded upon the supposition that what distinguishes life from nonlife is the capacity of a system to (1) self-organize and maintain self-organization for an extended period of time (O) and (2) undergo genetic-based reproduction (R). Precursors to O and R are found in the ancient writings of Aristotle, who identified "nutrition" and "reproduction" as the most basic characteristics of life and debated which one is more fundamental. Following Aristotle, most contemporary scientists and philosophers privilege one of these characteristics as more basic to life than the other. But even supposing, for the sake of argument, that these neo-Aristotelian characteristics are universal to life, it does not follow that they are fundamental to life. As we have seen, they could be symptoms of more basic but as yet unknown properties of life, in which case it would be a mistake to try to found a universal theory of life upon them. Likewise, most theories of the origin(s) of life are based upon variations on O or R (sometimes both), further underscoring the influence of Aristotle's ideas on modern attempts to answer fundamental questions about life. In short, the contemporary quest for answers to foundational biological questions is deeply rooted in Aristotelian assumptions. As Chapter 4 discusses, the Aristotelian character of biological thought may in part explain why we lack a truly general theory for even familiar life (let alone life considered generally); when presented with an allegedly general principle for life, many biologists delight in listing counterexamples.

As emphasized in this chapter, the most challenging aspect of characteristics O and R is making good sense of them causally. Unlike most contemporary scientific theories about natural phenomena, O and R characterize life functionally. Functional characterizations are highly abstract, which helps to explain why some ALife researchers are willing to countenance purely informational life implemented on digital computers; I will have more to say about computer life in Chapter 7. Functional properties are peculiar insofar as they are instantiated by an entity *before* the ends that functionally define them actually occur. That is, they are intrinsically goal directed. The concept of intrinsically goal-directed processes represents a problem for contemporary scientists and philosophers of science whose concept of causation holds that causes precede their effects without anticipating them. For this reason, functional capacities are rarely if ever taken as basic in science. Instead, they are treated as derivative, that is, as products of familiar, undirected, causal interactions among the components of systems exhibiting them. Mental states, which are *prima facie* intrinsically goal directed (e.g., my desiring a candy bar), provide an especially good illustration. Psychology treats them as functional properties. But even those who defend the autonomy of psychology – its irreducibility to neuroscience (and ultimately to chemistry and physics) – are loath to interpret mental states as requiring a truly novel form of physical causation. In the words of philosopher Jonathan Lowe (1993), "[t]hat mental phenomena are part of the natural, causal order of events is surely not to be denied" (p. 629). As with the case of life, the scientific challenge is showing how mental properties are generated by physicochemical processes through unproblematic causal interactions.

As discussed in Section 1.4, Darwin's theory of evolution by natural selection is commonly held up as showing how the ostensibly goal-directed characteristics of life can be explained in terms of linear chains of undirected causes and effects. The remarkable adaptedness of organisms to their environment is explicated as the product of a blind process of natural selection operating on heritable variations (mutations, recombination, and genetic drift) over long periods of time. The causal processes involved are complex, involving an enormous number of diverse interactions between organisms and their environments, but the causation involved is Humean. As discussed, however, even when fleshed out in light of twentieth century developments in biology, Darwin's theory cannot account for the transition from chemistry to biology. Admittedly, Darwin speculated (in an 1870 letter to his good friend the famous botanist Joseph Hooker) about life originating in a "warm little pond." But he says nothing about the nature of the causal processes involved.

Many advocates of the RNA World follow Manfred Eigen (1992) in invoking a principle of *molecular* natural selection to explain the development of the earliest living chemical systems from a population of (depending upon the version) self or collectively autocatalytic RNA oligomers. Primordial, self-catalytic RNA oligomers are characterized as "encoding" their catalytic efficiency as hereditary information in a neo-Darwinian sense (e.g., Alberts et al. 2002, Ch. 6). But the concept of encoding being invoked is so attenuated from a biological perspective that it is not clear that it can do the work required of it. A true encoding device converts information from one form to another. In the RNA World, however, there is no distinction between the encoding molecule and the encoded molecule; in the simplest version of the RNA World they are one and the same. In classical Darwinian evolution, in contrast, the distinction between the phenotype (metabolic, structural, and behavior characteristics) and the genotype of an organism is central to the selection process; the environment acts directly on the encoded phenotype and only indirectly on the genotype. In short, it is not clear that the specter of goal-directed causation can be expunged from the RNA World in a manner analogous to the way in which Darwin expunged it from advanced life.

But even supposing, for the sake of argument, that one could successfully explain the "evolution" of a primitive "rT riboorganism" (Benner 1987, 53) from a primordial collection of RNA oligomers in terms of a process of molecular natural selection, there is still the problem, discussed in Section 1.5.2, of the prebiotic assembly of the first catalytic RNA oligomers. Appealing to an even more general principle of chemical natural selection, which could just as easily be invoked by advocates of the SM World, will not work. For in the absence of greater specificity about the processes involved, such an appeal amounts to little more than invoking the (in essence, chemically miraculous) "spontaneous" self-assembly of a favored primordial biomolecule. The more sweeping and removed from biology a principle of natural selection, the less light it is able to shed upon the processes actually involved in the origins of life. This is perhaps most easily appreciated when one considers that some physicists have generalized natural selection even further, to a principle of *cosmological* natural selection (e.g., Smolen 1997), in efforts to explain the historical development of the universe. Such principles are too vague and sweeping to have genuine explanatory power for something as specific as the origin or nature of biological life. As Popper (1963) cautioned, the capacity of a scientific theory to explain everything is not strength but rather a weakness.

We are left in a quandary. Was Aristotle correct? Do answers to the most fundamental questions about life – its nature and origins – require recourse to a primitive form of goal-directed self-causation? Or, alternatively, does the difficulty in satisfactorily answering these questions lie in commitment to an unpropitious foundation of basic concepts and principles for theorizing about life? Put more provocatively, do we need to stop constructing theories of the nature and origin(s) of life on the basis of a theoretical framework founded more than two thousand years ago in ancient Greece?

During the last few decades the neo-Aristotelian concept of emergence has resurfaced in scientific and philosophical discussions about life. Whether an intelligible concept of emergence can be developed remains an interesting and open question, well worth pursuing. Some philosophers, most notably Kim (1992, 1999), contend that the concept is logically incoherent. Others (e.g., Bedau 1997; Humphreys 1996; Wimsatt 2007) are actively working on surmounting the logical problems identified by Kim and others. This book takes the road less traveled: It rejects a neo-Aristotelianism theoretical framework for reasoning about life. My choice is guided in part by skepticism about the coherence of the concept of goal-directed self-causation (tacitly appealed to in the concept of emergence) and in part by the history of science. Indeed, as Chapter 4 discusses, the rapid development and greatest empirical successes of the other natural sciences (most notably, chemistry and physics) occurred *after* the abandonment of Aristotelian concepts and principles. It strikes me as a bit scandalous that biology is the only natural science whose foundations are still dominated (albeit often tacitly) by neo-Aristotelian ideas. In any case, it seems clear that exploring alternative theoretical frameworks for theorizing about life is worthwhile, especially in light of some of the remarkable discoveries, discussed in Chapters 5, 6, 7, and 8, being made by molecular biologists and microbiologists.

2

Why Life Cannot Be Defined

2.1 Overview

Faced with the question 'what is life?' many scientists, and some philosophers, advance definitions of life. Defining life is especially popular among astrobiologists, many of whom are convinced that one cannot successfully search for truly novel forms of microbial life without a definition of life: How else will one recognize it if one encounters it? The extensive discussion of definitions of life in Part 2 ("Definition and nature of life") of the *CRC Handbook of Astrobiology* (Kolb 2018) provides a salient illustration of this attitude. Along the same lines, a recent version of the *NASA Astrobiology Strategy* (Hays 2015) contains a large section devoted to "Key Research Questions for Defining Life" (p. 145).[1] This chapter and the next explain why the scientific project of defining life is mistaken. Life is not the sort of thing that can be successfully defined. In truth, a definition of life is more likely to hinder than facilitate the discovery of novel forms of life.

Section 2.2 begins with a survey of the most influential contemporary definitions of life, specifically, thermodynamic, metabolic, evolutionary, and autopoietic definitions. As will become apparent, these definitions, which come in a variety of incarnations of varying degrees of abstractness (including purely informational), face robust counterexamples, and attempts to resolve them invariably generate new problems. It is thus hardly surprising that there is no consensus among scientists or philosophers on a definition of life. But, as Section 2.3 explains, incessant counterexamples to definitions of life are just the tip of the iceberg. It turns out that definitions are inherently incapable of answering 'what is' questions about natural categories.

---

[1] In addition to astrobiology, origins of life and artificial life researchers devote significant efforts to the issue of defining life in the belief that an understanding of how to create life, whether via natural processes, in a laboratory, or in the information structures of a computer, requires defining life; some origins of life researchers hold the opposite view, contending that one cannot "define" life without understanding how it originated. Chapters 5 and 7 explain why the emphasis on defining life among researchers into (respectively) the origins of life and artificial life is also mistaken.

Some categories of things seem natural in the sense that they would exist even if there had been no human beings to think about them. Water, bird, and gold provide good illustrations. Philosophers refer to groupings of this sort as "natural kinds."[2] Other groupings are non-natural in the sense that they depend upon human interests and concerns.[3] Examples of non-natural kinds include garbage, money, and art. The classification of a category of things as natural or non-natural may change in light of empirical and theoretical developments. On the basis of macroscopic mineralogical characteristics, such as hardness and color, jade was once classified as a natural kind. With the discovery that it includes molecularly distinct mineralogical compounds, jade was reclassified as a non-natural kind comprising two natural kinds (jadeite and nephrite). Similarly, in the 1970s, it was discovered that bacteria – which were viewed as a natural biological kind encompassing all prokaryotes (unicellular microorganisms lacking a membrane-bound nucleus) – consist of organisms differing biochemically and genetically from each other more than from single-celled eukaryotes. This discovery resulted in major revisions to the highest level of biological taxonomy, with the Kingdom Monera (at the time also known as Bacteria) being replaced by two new domains of life, Archaea and Bacteria, and all eukaryotes (protists, fungi, plants, and animals) being lumped together into a third domain, Eukarya. Whether life is a natural kind, like water, or a non-natural (human) kind, like jade, is an empirical question. The goal of this chapter is to show that *if* life is a natural kind (which is taken for granted by those championing definitions of life) attempts to define life are mistaken.

So, what exactly is a definition? Following earlier work (Cleland 2012; Cleland and Chyba 2002, 2007), Section 2.3.1 explores the logical character of definition, and concludes that definitions are the wrong tools for answering the question 'what is life?'; for a closely related argument against the project of defining life, see Machery (2012). As will become clear, definitions are concerned with language and concepts, as opposed to things in the world of nature. The best (most informative) definitions, which I dub "ideal," dissect the concept associated with a word or expression into sub-concepts providing necessary and sufficient conditions for the application of the term. Ideal definitions thus fully determine membership in

---

[2] As Chapter 4 discusses, naturalism about kinds does not presuppose that there is a unique way of carving nature at its joints. Nature is very complex. As John Dupré (1993) argues, there may be many highly plausible ways of dividing it into kinds. It does not follow, however, that these diverse systems of classification are equally good for the purpose of yielding scientifically fruitful, general theories, even supposing that they are useful for other purposes, such as cooking or horticulture. Furthermore, as we shall see, naturalism about the classifications of our most successful scientific theories does not (as some philosophers believe) imply the strong metaphysical doctrine of essentialism. Somewhat ironically, essentialism is closely bound up with the claim, rejected in this chapter, that natural kinds have defining properties (essences).

[3] Many philosophers who are hostile to essentialism reject this distinction, contending that all scientific categories are ultimately human kinds. This view seems highly problematic given the remarkable empirical successes of modern science; I will have more to say about this in Chapter 6.

the class of items designated by a term, which helps to explain their appeal. An astrobiologist could decisively settle the question of whether an extraterrestrial geochemical system is a living thing merely by consulting the "right" ideal definition of life.

As will become apparent, however, ideal definitions cannot live up to their promise. They are an ideal to which one can aspire but not something that can be actually achieved. The best definitions are subject to problems of vagueness, and plagued with counterexamples and borderline cases, which makes them especially poor tools for answering scientific questions about natural phenomena, such as life, that are not well understood. In this context, it is worth mentioning that it is not true, as some philosophers are fond of claiming, that scientists do not intend their definitions of life to be logically rigorous. Scientists not only seek necessary and sufficient conditions for life but also frequently get bogged down dealing with counterexamples and borderline cases. Sagan, Luisi, Kauffman, and Gánti, whose work is discussed in Section 2.2, provide especially salient illustrations because they either advance or explicitly discuss the need for supplying necessary and sufficient conditions when "defining" life.

The considerations sketched in Section 2.3.2 suggest that definitional approaches to explicating life are unlikely to be scientifically fruitful. For as discussed in Section 5.2, our current concept of life is based on a single example of life, and there are reasons for suspecting that it may be unrepresentative, which means that it is unlikely to shed much light on the general nature of life. The situation is even worse than this, however. In the second half of the twentieth century, philosopher Hilary Putnam (1973, 1975) mounted a devastating attack against definitional approaches to explicating natural kinds, showing that the meanings of natural kind terms cannot be captured by definitions.[4] Sections 2.3.3–2.3.4 explain this powerful but subtle argument, which revolutionized philosophical thought about the capacities and limitations of definition but is not well known among scientists. To cut a long story short, natural kind terms *cannot* be defined. As Chapter 3 argues, what is needed to answer the question 'what is life?' is a universal *theory* of life. Yet, as Chapter 5 explains, we currently lack such a theory: To formulate one we need samples of life as we don't know it. And this presents astrobiologists with a potential dilemma: How will they recognize truly novel forms of life should they be fortunate enough to encounter them? A solution to this dilemma is supplied in Chapter 8, which develops a novel strategy for searching for life as we don't know it in the absence of either a definition or a truly general theory of life.

---

[4] Saul Kripke (1972) developed a similar line of argument but, while he applied it to natural kind terms, his central focus was proper names. Putnam's work provides a far more extensive and sustained attack on the notion that natural kind terms can be defined.

2.2 Popular Definitions of Life

The history of attempts to define life is very long. As mentioned in Section 1.5.1, most scholars believe that it dates back to the writings of Aristotle. Scientific interest in defining life was rekindled in the latter part of the twentieth century by astronomer Carl Sagan's now classic essay in *Encyclopaedia Britannica* (Sagan 1970). Writing during a period of rapid advances in molecular biology, computer technology, and space science, Sagan reviewed the then most widely accepted scientific definitions of life – physiological, metabolic, biochemical, thermodynamic, and Darwinian (which he somewhat misleadingly called "genetic") – and showed that each faces intractable counterexamples, either including things that are clearly not alive (e.g., candle flames and quartz crystals) or excluding things (e.g., dormant seeds and spores, and mules) that are clearly biological. Sagan's essay produced a flurry of new activity on definitions of life. Some of the definitions that he discussed (e.g., physiological) were abandoned. Others were reformulated in hopes of undermining the strongest counterexamples advanced against them. New definitions of life, for example, autopoietic, were also introduced. The upshot is that we currently have an impressively large and diverse collection of definitions of life; see Pályi et al. (2002), Popa (2004, Appendix B), and Trifonov (2011), for many examples.

Yet despite these efforts, there is no widely accepted definition of life (Chyba and McDonald 1995; Cleland and Chyba 2002, 2007; Tsokolov 2009). Although reworked and refined in light of contemporary scientific advances, the most popular contemporary definitions (thermodynamic, metabolic, evolutionary, and autopoietic) still face serious problems, ranging from robust counterexamples to being too *ad hoc*, i.e., designed expressly for the purpose of fortifying a favored view of the nature of life against what is *prima facie* a devastating counterexample. A brief survey of these definitions and their problems is illuminating.

2.2.1 Thermodynamic Definitions

Thermodynamic definitions of life are popular among biophysicists. Commonly associated with the writings of physicist Erwin Schrödinger (1944), a basic thermodynamic definition might characterize a living system as one that creates and maintains local order (organismal bodies and ecosystems) by extracting energy ("drinking order") from the environment over an extended period of time. As Sagan noted, however, many inanimate physical systems create order by extracting energy from their environment. Mineral crystals, such as quartz or diamond, provide good illustrations.

More nuanced thermodynamic definitions of life have been developed in the hopes of avoiding counterexamples of this sort, for example, Hoelzer et al. (2006),

Lineweaver (2006), Schneider and Kay (1994), and Schulze-Makuch and Irwin (2008). But they are not free of troubling counterexamples either. In this light, Stuart Kauffman (e.g., 2000, 14–22), who is convinced that life is a novel thermodynamic phenomenon, makes a radical suggestion: A new "fourth law of thermodynamics," applying to complex, far from equilibrium, chemical reaction systems that are "self-constructing" is required for understanding life. Kauffman develops a concept of "autonomous agency" based on this idea and attempts to define life in terms of it; see, for example, Kauffman (2003) and Kauffman and Clayton (2006). From the perspective of this book, one of the more interesting aspects of his work is the resurrection of the old Aristotelian notion of self-causation (discussed in Chapter 1) in a more contemporary scientific guise.

The details of Kauffman's definition are unimportant for our purposes. What is important is that his reasons for proposing a new law of thermodynamics – namely, the inability to explicate life in terms of established thermodynamic considerations – should count *against* the viability of a thermodynamic definition of life. To propose a vaguely described new law of thermodynamics *solely* for the purpose of salvaging a thermodynamic account of life is *ad hoc*. In this context, it is important to keep in mind that there is nothing *ad hoc* about pursuing Kauffman's conjecture as a *research project* – formulating more precisely described candidates for a fifth law of thermodynamics and exploring a diversity of physicochemical systems for empirical support. This highlights an important difference, discussed in much greater detail in Chapter 3, between definition-based and theory-based approaches to universal biology.

2.2.2 Metabolic Definitions

As discussed in Chapter 1 (Section 1.5.1), metabolic and thermodynamic definitions of life are closely related, the primary difference between them being that the former focus on the flow of material (versus energy, interpreted statistically as local order) between a physical system and its environment. Sagan's nuts and bolts version of a metabolic definition characterizes life as "an object with a definite boundary, continually exchanging some of its materials with its surroundings, but without altering its general properties at least over some period of time" (Sagan 1970). As Sagan observes, however, this definition faces compelling counterexamples. It includes nonliving phenomena such as candle flames, which have a definite shape and maintain themselves by continually exchanging material with their environment – extracting oxygen and combining it with wax, and releasing carbon dioxide and water in the process – and excludes viable seeds and spores, which may lie dormant for hundreds or even thousands of years before germinating.

More sophisticated chemical metabolic definitions have been developed in efforts to ward off counterexamples of these sorts, for example, Dyson (1999), Feinberg and Shapiro (1980), Hazen (2005). Some restrict the possible chemistries of metabolic processes to organic (carbon-containing) molecules. Basing a definition of life too closely on the chemistry of familiar Earth life is problematic, however, insofar as it excludes forms of life that are conceptually and perhaps even scientifically plausible. The Horta of *Star Trek* fame provides an especially colorful illustration. In the episode *The Devil in the Dark*, the Horta, an amoeba-like blob of rock, confounds Captain Kirk and his crew by burrowing through solid rock and behaving in a life-like manner. The Horta turns out to be a truly alien, chemical form of life (based on silicon rather than carbon); Mr. Spock discovers, via a Vulcan mind meld (of course), that it is a "mother" defending her eggs from miners.

It is of course an open question whether life forms differing this radically from Earth life in basic chemical composition are biologically possible, but surely one does not want to rule them out *merely by definition.* Besides, there are theoretical reasons for thinking that silicon-based life forms such as the Horta may not be as scientifically implausible as sometimes supposed. Schulze-Makuch and Irwin (2008) argue that under physical and chemical conditions very different from those on Earth (viz., little free oxygen or liquid water, and very low temperatures), silicon may be able to form longer and more complex molecular structures; molecular size and complexity have long been thought to be a requirement for life (Pace 2001). They conjecture that there might even be silicon-based life forms on Titan, a moon of Saturn. As biochemist William Bains observes, however, Mr. Spock's experience with Titanians would not be very pleasant. Even a small whiff of their breath would stink to high heaven, and if a Titanian were beamed aboard the starship Enterprise it would boil, burst into flames, and kill everyone in the transporter room (Heward 2010).

A less extreme but far more worrisome illustration of how a metabolic definition of life based too closely on the biochemistry of familiar Earth life could hinder the recognition of extraterrestrial life is provided by the Viking spacecraft missions to Mars in the 1970s. The purpose was to search for microbial life. To this day they remain the only dedicated in situ search for extraterrestrial life that has actually been conducted.[5]

---

[5] It is important to distinguish searching for habitable environments from searching for life. A number of missions have explored the habitability of planets and moons within our solar system by looking for evidence of liquid water, an atmosphere with the right chemical composition, etc. But the presence of these conditions does not provide direct evidence of life (past or present). For the latter, one needs to look for biosignatures (telltale traces of life). Two missions to be launched in 2020 have instrument packages dedicated to looking for biosignatures: NASA's 2020 SHERLOC and the second part of the ExoMars mission, sponsored jointly by the

The Viking biology package consisted of three experiments. As the Biology Team Leader, Harold Klein, recounts, these experiments were designed to test a number of "metabolic models of possible Martian biology" (Klein 1978b, 157) centered "... on the assumption that Martian biology would be based on carbon chemistry" (Klein 1991, 259). In other words, they were implicitly based on a chemical-metabolic definition of life. The basic idea was to search for evidence of Earth-like microbial metabolism. A miniature gas chromatograph mass spectrometer (GCMS) for detecting organic molecules was also included.

The experiment that produced the most baffling results was the Labeled Release (LR) experiment; see Section 8.2 for a more detailed discussion. In the LR experiment Martian soil was robotically introduced into a test chamber and injected with a carbohydrate solution, labeled with radioactive carbon-14, that is readily metabolizable by a wide range of culturable microorganisms. To the delight of the Viking biology team the soil started evolving $^{14}CO_2$ – just what one would expect from Earth microorganisms. When a sample of Martian soil was heated to 160 °C for three hours, more than enough to kill any known microorganisms, and then inoculated with the nutrient solution, no $^{14}CO_2$ gas was released, strongly suggesting that the initial reaction was biological. When a sample of Martian soil, that had evolved $^{14}CO_2$ gas after an initial inoculation, was given a second helping of nutrients, however, the anticipated burst of new activity from hungry Martian microbes failed to occur. Even more mysteriously, $^{14}CO_2$ left over from the initial reaction began disappearing (Levin and Straat 1977). The biology team was stunned. They could not explain these results in terms of any known chemical process, biological or nonbiological. In the end, they concluded that it must not have been biological on the basis of the anomalous (for Earth microbes) results and, most importantly, the failure of the Viking GCMS to find any organic molecules in the soil to its limits of detection with sample heating up to 500 °C (Biemann et al. 1976, 1977).

A consensus was quickly reached by the majority of the astrobiology community that the mysterious results of the LR experiment were produced by some sort of inorganic oxidant in the Martian soil. Various possibilities, including hydrogen peroxide, superoxides, and exotic oxidation states of iron, and, most recently, perchlorate, have been explored over the years. None are able to fully explain all the results of the LR experiment; see Section 8.2. In addition, it has been shown that the Viking GCMS, which was the linchpin in the original argument against biology, could not have detected as many as 10^6 bacterial cells per gram of soil, more than are found in some samples of Antarctic soil (Bada 2001; Glavin et al.

European Space Agency (ESA) and the Russian space agency (Roscosmos). The latter will be the first missions dedicated to searching for life since the Viking missions.

2001; Navarro-Gonzáles et al. 2006). Yet NASA has not placed a high priority on examining the Martian surface for the presence of a powerful inorganic oxidant. Why? Because (as a NASA scientist involved in the mission informed me) they were looking for life and what they found was not life. In other words, the definition undergirding the Viking life detection instrument package was not satisfied, and hence: no life. This attitude illustrates both the potential power of an accepted definition of life to blind scientists to alternative forms of life and the dangers of basing a definition of life too closely on the biochemistry of familiar Earth life. In short, we might not recognize an alternative form of life if we search for it with an inadequate definition of life.[6]

2.2.3 Evolutionary Definitions

Evolutionary accounts currently dominate scientific and philosophical thought about the nature of life. Most evolutionary definitions of life are founded upon Darwin's theory of evolution by natural selection. A few are more inclusive, postulating either a more general evolutionary mechanism, subsuming natural selection as a special case, for example, Bedau's (1998) "supple adaptation," or identifying natural selection as one of several evolutionary mechanisms, for example, Jablonka and Lamb (2005), who argue that epigenetic inheritance represents a neo-Lamarckian mechanism of evolution. Our focus in this section, however, is on Darwinian definitions.

The most abstract versions of the Darwinian definition of life follow Dawkins (1983) and Dennett (1995) in holding not only that the capacity to undergo evolution by natural selection is universal to life ("universal Darwinism") but also that it is independent of the material stuff that instantiates life. Darwinian definitions of this type are especially popular among artificial life (ALife) researchers, who frequently view evolution by natural selection as an algorithmic process (reducible to a system of equations). As an illustration, Rasmussen (e.g., 1992) and Ray (2009) contend that a sufficiently complex computer simulation exhibiting Darwinian evolution would be truly alive. Indeed, Ray contends that he has already created digital organisms in the informational structure of a computer. On Ramussen's and Ray's version of the Darwinian definition, it is not the computer that is alive but the computational processes themselves; the computer's role is analogous to that of glassware that may someday be used in the artificial synthesis of chemical life in a laboratory.

Most biologists and biochemists, however, find the idea of an immaterial computer simulation of life actually *being* alive counterintuitive. Few biochemists believe that a computer simulation of biochemistry is biochemistry itself.

---

[6] I am not of course claiming that the Viking spacecrafts found Martian life; on my view, the jury is still out.

A computer simulation of photosynthesis does not produce authentic carbohydrates (chemical compounds of carbon, hydrogen, and oxygen); at best, it may be said to yield simulated carbohydrates. Analogously why would one think that a computer simulation of life is actually alive, as opposed to merely simulating life? The point is that the only unequivocal example of life available to us is familiar biological Earth life, and it is chemical. One cannot, of course, rule out someday discovering that the essential characteristics of life are independent of the stuff that realizes them. It is important to keep in mind, however, that there currently exists no empirical support whatsoever for this claim. All that a computer simulation of life establishes is that some of the high level, functional characteristics of life can be successfully emulated on a computer, which is hardly surprising because computers can simulate the purely structural and functional characteristics of virtually any physical system, for example, bridges, landslides, hurricanes, and the human circulatory system. In this context, it is especially important not to be seduced by the starry-eyed claim that computers are on the threshold of providing us with the first truly alien creatures. For to recapitulate, the claim that a computer simulation of a living thing is itself alive rests upon the dubious assumption that the basic properties of life are independent of their physical embodiment, and this assumption is exactly what is at issue. We will return to this issue in Chapter 7.

In light of these concerns, many biologists and biochemists explicitly restrict Darwinian definitions of life to chemical systems. The "chemical Darwinian definition," which became NASA's official "working definition" of life in the 1990s, is the prototype and remains popular (NRC 2007). According to the chemical Darwinian definition "life is a self-sustained chemical system capable of undergoing Darwinian evolution" (Joyce 1994). As Joyce explains,

the notion of Darwinian evolution subsumes the processes of self-reproduction, material continuity over a historical lineage, genetic variation, and natural selection. The requirement that the system be self-sustaining refers to the fact that living systems contain all the genetic information necessary for their own constant production (i.e., metabolism).

(pp. xi–xii)

Although it is often interpreted as such, the chemical Darwinian definition is not restricted to organic molecules. As a result it is open to the possibility of silicon-based life on Titan, or perhaps even sulfur-based life in the sulfuric acid clouds of Venus (Grinspoon 2003, 283–286). As the next subsection discusses, however, it is difficult to design instrument packages for detecting organic (let alone inorganic) forms of extraterrestrial life on the basis of the chemical Darwinian definition. This undermines its utility for searching for extraterrestrial life. But this is only a practical problem. The definition faces much more serious problems.

The chemical Darwinian definition cannot accommodate the possibility that cellular life on Earth passed through a stage in which it could reproduce in a rudimentary fashion but could not yet engage in Darwinian evolution. According to some metabolism-first models of the origin of life – see Section 1.5.2 and Section 5.4 – the earliest forms of Earth life were autocatalytic chemical reaction systems consisting of networks of encapsulated, small organic molecules, primarily amino acids, cofactors, and peptides. Protocells in these reaction networks could divide but, lacking nucleic acid-type genetic systems, were not yet capable of Darwinian evolution. Furthermore, some advocates of the RNA World (a genes-first theory of the origins of life) hold a similar view. Carl Woese (1998; Vestigian et al. 2006), for instance, points out that Darwinian evolution requires a very complex genetic apparatus (with a universal genetic code coupled to a very precise mechanism of translation). It is unlikely that a hereditary apparatus of this complexity emerged in one fell swoop with the first protocells. Woese conjectures that cellular life, as we know it today, was preceded by collections of cellular entities ("progenotes") capable of primitive cell division (by pinching off) but with genetic systems (consisting of small mobile genetic elements subject to rampant mutation and haphazard transfer from one progenote to another) too inaccurate to support Darwinian evolution. On the chemical Darwinian definition of life, neither progenotes nor their communities qualify as living things. Yet if such entities existed (or perhaps still exist on some other world) it would surely be a mistake to deny them the status of living things solely because they fail to satisfy a preferred, preconceived, definition of life.[7]

Another potential difficulty, which the original version of the chemical Darwinian definition shares with many other evolutionary definitions of life, concerns the status of individual organisms, such as a bacterium, cat, or tree, as living things. Biological evolution occurs at the level of populations of organisms. Individual organisms (whether fertile and reproducing, fertile and nonreproducing, sterile members of fertile species, or sterile hybrids) do not *themselves* evolve. Most champions of evolutionary definitions recognize that individual organisms provide the paradigm for a living thing and hence that an inability to classify individual organisms as living things undermines the conceptual plausibility of a definition of life. A strategy for circumventing this difficulty is to distinguish life from the entities (individual organisms) that participate in it. Philosopher Mark Bedau (1998), for example, defines life as an evolutionary process ("supple adaptation") in which individual organisms (fertile as well as sterile) participate as derivative

---

[7] In this context, it is worth noting that one cannot successfully defend the chemical Darwinian definition on the grounds that a genetic system is required for an autocatalytic chemical network to achieve the level of organized complexity required of a primitive metabolism. For as discussed in Section 5.4.2, a genes-first theory of the *origins* of life is fully compatible with a metabolic account of the *nature* of life.

("secondary") forms of life. Similarly, biochemist Steve Benner (2010) distinguishes "life" as an evolutionary (population-level) system from its "parts," which include individual organisms that are said to "be alive." As Benner puts it, "one rabbit may be alive even though he or she is not life" (p. 1021).

Efforts to shore up evolutionary definitions of life in this manner are not problem free, however. Consider the very real (but thus far hypothetical) possibility of artificially synthesizing a microorganism from scratch (bottles of nonliving chemicals) that cannot reproduce and is unable to hybridize (swap genes) with other microorganisms. Such microbes would not qualify as alive on either Bedau's or Benner's definition of life; it does not participate in an evolving (population-level) system.

A well-worn strategy for resolving difficulties of this sort is to modify the definition so that the troublesome entities are included within its scope. One might, for example, stretch the concept of evolution to include being constructed artificially by a creature that itself is the product of Darwinian evolution. This does not fully solve the problem, however, because we cannot rule out the possibility of designing robots to synthesize microorganisms of the sort in question. One could, of course, resolve this difficulty by stretching the concept of evolution even further to include microorganisms synthesized by robots, constructed by robots, …, constructed by creatures that are the products of Darwinian evolution. But now we have another problem, namely, the concept of evolution has been stretched so far beyond the original biological notion that it is not clear that it qualifies as Darwinian anymore. The central problem should be obvious. Whatever strategy one comes up with – including biting the bullet and declaring that the artificial microbes concerned are not alive – is extremely *ad hoc*, that is, motivated solely by a desire to protect a preferred definition from what would otherwise be intractable counterexamples. Any definition of life can always be salvaged in this contrived manner but at a high cost, namely, undermining its capacity to shed light on the nature of life.

2.2.4 Defining Life as Self-Organized Complexity

In recent years, newer definitions focusing on the self-organizing properties of organisms have been developed. The most influential are autopoietic definitions (Luisi 1993; Maturana and Varela 1973) and Gánti's 1971 chemoton model of life (see Gánti 2003). Kauffman's definition of life in terms of autonomous agency should also be included. For as Luisi observes, all three modes of definition "… share a common view about minimal life as a distributed emergent property based on an organized network of reactions and/or processes" (Luisi 2003, 53). They differ primarily on how they understand self-organized complexity.

Kauffman cashes it out in terms of physics, the product of a heretofore unrecognized, new law of thermodynamics. Autopoietic definitions and the chemoton model, in contrast, regard it as a logical property of a system (Luisi 2003, 53). Damiano and Luisi (2010) define life as "... a system capable of self-production and self-maintenance through a regenerative network of processes which takes place within a boundary of its own making and regenerates itself through cognitive or adaptive interactions with the medium" (p. 149). Gánti (2003) develops a formal model of life, as "fluid automata," described by a system of equations.

The most obvious difficulty with all three approaches is that the vague concepts of self-organization (e.g., "emergence," "self-referentiality," "self-sustainability," and "auto-maintenance") invoked to explain life are treated as primitive, raising the hoary old problem, discussed in Chapter 1, of goal-directed self-causation. The basic idea seems to be that at a certain level of organization, novel functional properties spontaneously arise, as a matter of either fundamental physics or logic, rendering the system capable of developing increasingly greater organized complexity. Advocates of these definitions seem to believe that the idea of novel properties arising spontaneously at higher levels of organization is an asset, but as pointed out in Chapter 1, it is actually a weakness. In truth, these approaches to explicating the nature of life have much in common with older views such as vitalism and British emergentism, with the downside that less attention is paid to the physical and metaphysical problems posed by the notion of goal-directed self-causation.

In addition to the problem of making good sense of goal-directed self-causation, autopoietic definitions, which have received more attention than the other two, are notable for their permissiveness. They classify many systems that would otherwise be considered paradigmatically nonliving (and hence, compelling counterexamples) as living things. Some provocative illustrations are whole ecosystems, human institutions (such as corporations), and even Earth itself (Lovelock 2000; Margulis and Sagan 1995)[8]; computer simulations are also commonly included on the grounds that the difference between life and nonlife is merely a matter of how the components of a system are organized and integrated. That a proposed definition admits so many *prima facie* nonliving entities (seeming counterexamples) as living individuals would normally count against it. Turning the tables and insisting that they are living things *just because* a favorite definition says so is logically problematic.

As earlier, it is important to keep in mind that there is nothing problematic about attempting to develop a coherent concept of autopoiesis and exploring its potential for shedding light on unambiguous cases of life. Furthermore, as Chapter 4

---

[8] The claim that the Earth itself is a living individual is known as the Gaia hypothesis; it was first proposed by Lovelock (1972) and codeveloped with Lynn Margulis (Lovelock and Margulis 1974).

discusses, research within the context of an empirically successful scientific theory sometimes changes classification schemes in significant ways. It is thus foolish to insist that our current classifications of entities as living and nonliving are the final word. But such a project is very different from formulating a logically rigorous definition that (in virtue of supplying necessary and sufficient conditions for life) fixes what is and is not a living thing in one fell swoop. The point is, as discussed below, the best definitions are brittle. As Chapters 3 and 4 discuss, one of the strengths of scientific theories is the open-endedness of their classification schemes, which allows them to evolve in response to new empirical discoveries and theoretical insights.

2.3 The Problem with Definitions

That the most popular contemporary definitions of life are defective does not of course establish that no definition of life could be successful. Efforts to formulate more satisfactory definitions of life are ongoing. Some attempt to resolve problems with earlier definitions by combining the collective (evolutionary) and individualistic (metabolic) aspects of life, instead of trying to "reduce" one to the other; see, for example, Ruiz-Mirazo et al. (2004). Nevertheless, it is striking that, even with the remarkable advances in biology and biochemistry over the past couple of centuries, we still lack a consensus on a definition of life. This lack of consensus over such a long period of time suggests that there may be something wrong with the project of defining life. To successfully undermine the project of defining life, however, it is not enough to catalog difficulties with particular definitions. One needs a more general argument – one showing that definitions are *inherently* incapable of performing the task required of them. Drawing upon groundbreaking work by Putnam (1973, 1975) on the meaning and reference of natural kind terms, the remainder of this chapter develops such an argument. In order to fully appreciate its strength, however, one first needs a good understanding of the logical character of definition.

2.3.1 *Logical Character of Definition*

The stereotypical example of a definition familiar from introductory logic classes is: 'Bachelor' means unmarried human male. As the use of single quotation marks indicates, this definition explains the *meaning* of a *word* ('bachelor'), as opposed to talking about the things (actual bachelors) that the word designates.[9] The term

---

[9] Philosophers use single quotes around a word to indicate that one is talking about the *word*, as opposed to the *things* that it designates. Failure to do so represents a violation of the *use-mention distinction*, which is crucial in logic and philosophy of language. As an example the claim that 'bachelor' has eight letters is not about the men designated by the word 'bachelor'. In contrast the claim that bachelors are unhappy is not

being defined is known as the *definiendum* and the term doing the defining is known as the *definiens*. There is a long tradition in philosophy of identifying meanings with concepts – with what is known or intellectually grasped in understanding a word or linguistic expression. On this tradition, meanings are, as Hilary Putnam (1975) famously put it, "in the head." In other words, meanings are not identified with the items designated by a term but rather with the concept that we associate with the term, which underscores a crucial and often overlooked point in discussions of definitions of life: Definitions, in the standard logical sense, are not concerned with mind-independent things in the world of nature. They are concerned with language and thought.

Definitions come in a variety of different forms, each designed to meet a special need for communicating linguistic meaning (see, e.g., Audi 1995). This section focuses on those commonly used by scientists and philosophers to define 'life', namely, operational definitions and (what I have dubbed) "ideal definitions"; Section 2.3.3 discusses stipulative definitions, which were popular among philosophers of science during the first half of the twentieth century. For purposes of orientation, however, it is useful to begin with the most familiar form of definition: Dictionary or (more technically) lexical definitions.

Lexical definitions report on standard meanings of terms in natural languages such as English. Unfortunately, lexical definitions are usually circular, defining words in terms of close cognates. The circularity of lexical definitions deprives them of utility for a speaker who has not mastered the pertinent concepts or who is unfamiliar with the terms being used in the definiens. A good illustration is defining 'line' to mean linear path; 'line' is the grammatical root of 'linear'. The circularity may be subtler, however, as is the case when one defines 'cause' to mean something that produces an effect; the meanings of 'cause' and 'effect' are so tightly intertwined that it is unlikely that someone unfamiliar with the former will be familiar with the latter. Lexical definitions are thus poorly suited for expanding our understanding of the items designated by linguistic expressions. At best, a lexical definition could reveal what 'life' means to English speaking people. But when a natural scientist asks about the nature of life she is not interested in what the term means to twentieth century English speakers. She wants to know what life truly is, regardless of how competent speakers of the English language happen to

---

about the word 'bachelor'; words are not the sort of things that can be unhappy. The use of single quotes around a word in standard logical discussions of definition signals that the word is being talked about (*mentioned*). A lack of single quotes indicates that one is *using* the word to talk about the things to which the word refers. The use of single quotes in standard logical discussions of definitions signals that definitions are directly concerned with words (in this case, their meanings, as opposed to number of letters) and only indirectly concerned with the things that the words designate. Failure to honor the use-mention distinction can result in serious logical confusions. To reinforce the reader's awareness of this risk, I use single quotes in the remainder of this section to indicate when I am talking about a word, as opposed to the things that it designates. I subsequently drop the use of single quotes except in cases where confusion might result.

use the word. It should therefore come as no surprise that philosophers and scientists do not cite the *Oxford English Dictionary* when constructing "definitions" of life.

Operational definitions are very popular among scientists, and hence it should come as no surprise that they frequently advocate their use for defining 'life', for example, Chao (2000), Chyba and McDonald (1995), Conrad and Nealson (2001), McKay (1994). Like their close cousins, ostensive definitions, operational definitions use representative examples to indicate the meanings of terms. Instead of gestures (pointing to examples) or lists (naming examples), however, they use procedures. Defining 'acid' to mean something that turns litmus paper red provides a salient illustration. It specifies a method ("defining test") for determining whether a sample of an unknown liquid is an acid.

As discussed in the last section, Darwinian definitions of life are currently in vogue among biologists. They are difficult to operationalize, however, because evolution by natural selection takes place over many generations, making it difficult to detect in large multicellular creatures (plants and animals), which typically reproduce slowly and live for long periods of time. Microbes are another story, however. Lin Chao (2000) has proposed an ingenuous operational definition for detecting Darwinian evolution in microbes. Because Earth microbes reproduce rapidly, significant increases in fitness (measured in terms of population parameters such as growth rate, total density, and lag time) can be detected in laboratory cultures within a period of a few weeks to a month (Atwood et al. 1951; Lenski et al. 1991). Astrobiologists are especially interested in operational definitions of life that apply to microbes because, as discussed in Chapter 5, microbial forms of life are thought to be far more common in the universe than large multicellular organisms. Chao's operational Darwinian definition is specifically designed for developing instrument packages for detecting extraterrestrial microbes on planets and moons in our solar system.

Nevertheless, despite being billed as an operational definition, Chao's definition is not very practical. It presumes that the environmental and nutritional needs of extraterrestrial microbes will be very similar to those found on Earth and moreover that they will reproduce at a similar (rapid) rate. As discussed in Section 5.3.1, however, less than 1% of Earth microorganisms have been cultured, presumably because their environmental or nutritional needs are not being met. Extraterrestrial microbes are even more likely to have poorly understood and unknown needs of this sort, and hence to be even more difficult to culture. Besides, even supposing that they could be cultured, alien microbes might reproduce far more slowly than Earth microorganisms, too slowly to detect an improvement in the capacity to use a nutrient solution within the time frame of an in situ robotic experiment on another world.

The biggest drawback of operational definitions, however, is that they cannot provide compelling answers to 'what is' questions about natural categories. A good operational definition specifies a procedure for deciding whether an item happens to be a member of the category being defined. But it does not specify what unifies the items as members of that category. As an illustration, litmus paper allows one to identify that a particular sample of liquid is an acid but it does not tell us anything about the *nature* of acidity.

It is important to distinguish an operational definition *simpliciter* from an operational definition based on a successful scientific theory. In the latter case, one is able to answer the 'what is' question but the work is done by the theory, not the definition. The claim that an acid is "a substance that provides hydrogen cations (H^+) when dissolved in water" (McMurry and Fay 2001, 56) provides a fairly compelling chemical answer to the question 'what is an acid?' But this answer does not comprise an operational definition of acidity; it does not specify an actual procedure for identifying which substances are acids and which are not. Given enough information about the chemical composition of litmus paper, in the context of the chemical theory of acidity, however, one could infer that litmus paper is the sort of thing that turns red when immersed in an acid. It is for this reason that litmus paper is considered a good "test" for acidity. In the absence of an adequate theory of acidity, however, operationally defining 'acid' as something that turns litmus paper red is not very compelling. Analogously, an operational definition of life that is not founded upon an empirically well-grounded, general theory of life is unlikely to tell us much about the nature of life and, even worse, is likely to badly mislead us when it comes to recognizing truly alien forms of life for what they represent. To appreciate this point, one need only reflect on the official NASA interpretation of the equivocal results of the Viking life detection experiments. To rework an old saw: We won't recognize it when we see it unless it closely resembles what we are familiar with.

The most promising candidates for defining 'life' are (philosophically) *ideal* definitions; these definitions are sometimes called "full" or "complete" but because these terms are also used by philosophers to designate more fine-grained distinctions than concern us, I have elected to use a different term. Ideal definitions define terms by analyzing the concepts associated with them into logical conjunctions of properties. The conjunction of properties supplies necessary and sufficient conditions (an identifying description) for the application of the term being defined, and hence completely determines whether something falls into the class of items designated by it; the latter is known as the *extension* of the term.

The informative power of an ideal definition depends upon the terms in the definiens being better understood than the term being defined. As an illustration, the definition of 'bachelor', with which we began this discussion, provides a complete analysis of the concept of bachelor in terms of familiar concepts, viz., being unmarried, human, and male. The requirement that the terms in the definiens be better understood than the term being defined highlights a serious but often overlooked defect of autopoietic definitions of life. The concepts of self-reference and emergence are obscure and poorly understood, which seriously undermines the explanatory power of an autopoietic definition of life. It is always a mistake to define a poorly understood word in terms of an equally poorly or even less well understood expression.

Ideal definitions represent a philosophical aspiration that alas is rarely achieved in practice, which explains my choice of the term 'ideal' to characterize them. Most purportedly ideal definitions face borderline cases in which it is uncertain whether something satisfies the conjunction of predicates supplied by the definiens. Our definition of 'bachelor' is no exception. A ten-year-old boy qualifies as a bachelor because he is unmarried, human, and male. One could resolve this difficulty by adding 'adult' to the definiens, but doing so will not completely eliminate borderline cases, for example, the status of a young man who will turn eighteen in a few days, or for that matter, the Pope.

Difficulties with borderline cases reflect a fundamental feature of natural languages. Most general terms are vague; their reference is not fully determined. The famed sorites paradoxes provide good illustrations: Consider the problem of trying to draw a precise distinction between men who are bald and men who are not. One cannot do it in terms of the number of hairs on a man's head; our concept of baldness is not this precise. But it does not follow that there is no difference between being bald and being not-bald. Although definitions specifying conditions that are truly necessary and sufficient for the application of a term are rare, one can usually construct fairly satisfactory approximations. The important issue is not whether a definition can eliminate all borderline cases but whether it gets the indisputable cases right, for example, men sporting thick heads of hair in the case of 'not-bald', and forty-year-old unmarried men in the case of 'bachelor'. As a consequence, it is a mistake to think that viruses and prions pose serious threats to various definitions of 'life'. As far as our current concept of life is concerned, their status as living or nonliving is genuinely unclear. They represent *bona fide* borderline cases, and hence it is only to be expected that definitions of 'life' have trouble classifying them. In contrast, most of the counterexamples cited by Sagan (1970) do not involve borderline cases. They involve cases (e.g., candle flames, quartz crystals, and viable bacterial spores) that lie clearly on one side of the boundary between the living and the nonliving. These are the cases that matter for purposes of definition. They provide authentic counterexamples to definitions of 'life'. Any definition that fails to properly classify them is seriously flawed.

2.3.2 *Limits of Definition*

Within the constraints of vagueness, ideal definitions provide excellent answers to at least some 'what is' questions. Indeed, it is hard to imagine a more informative answer to 'What is a bachelor?' than "an unmarried, [adult] human male." The question is can we formulate an ideal definition of 'life' that provides a similarly satisfying answer to the question 'what is life?' As explained below, the answer is "no."

The implausibility of explicating the nature of life by means of a definition is best appreciated in the context of a somewhat contrived example from the history of science: Suppose that a seventeenth century scientist attempted to answer the question 'What is water?' by defining 'water'. This is before the advent of molecular theory, whose foundations were laid in the late eighteenth century by Antoine Lavoisier. As a consequence, our hypothetical scientist knows nothing of molecules. His knowledge of water is limited to phenomenal (sensible) properties such as being wet, transparent, tasteless, odorless, and a good solvent. Unfortunately, many of the substances identified by his contemporaries as water lack one or more of these properties. Muddy water is not transparent, salty water is not tasteless, and brackish water is not odorless. It is difficult for us, steeped in twenty-first century chemistry, to fully appreciate the dilemma that our seventeenth century scientist faces in selecting one or more of the phenomenal properties associated with water as fundamental to it. The great sixteenth century polymath Leonardo da Vinci, who had a life-long interest in the nature of water, expressed it well:

And so it [water] is sometimes sharp and sometimes strong, sometimes acid and sometimes bitter, sometimes sweet and sometimes thick or thin, sometimes it is seen bringing hurt or pestilence, sometimes health-giving, sometimes poisonous. So one would say that it suffers change into as many natures as are the different places through which it passes. And as the mirror changes with the colour of its object so it changes with the nature of the place through which it passes: health-giving, noisome, laxative, astringent, sulphurous, salt, incarnadined, mournful, raging, angry, red, yellow, green, black, blue, greasy, fat, thin. . .
(Il Codice Arundel, No. 263, fol. 57r; trans. MacCurdy 2003, 734)[10]

What is striking about this passage is the confusion evinced by da Vinci over the nature of water. He is utterly perplexed. Different samples of water exhibit such a wide variety of characteristics (sharpness, thickness, bitterness, sweetness, saltiness, greenness, greasiness, etc.) that he despairs of finding something that they all have in common.

Early modern alchemist-chymists were impressed by water's remarkable powers as a solvent. As a consequence, they treated solvency as a fundamental property of water and classified solutions of nitric acid and mixtures of nitric acid and hydrochloric acid (both of which are even better solvents than water, and share

---

[10] I am grateful to Christopher Chyba for discovering this marvelous quotation.

some of its other sensible properties, e.g., being transparent and a liquid) as water; it is not an accident that the former was called *aqua fortis* (strong water) and the latter *aqua regia* (royal water) (Roberts 1994). But as we now know, they were wrong. What distinguishes water from all other chemical substances is a unique molecular composition, H_2O. Being composed of molecules of H_2O is what salty water, muddy water, brackish water, and distilled water have in common, despite their phenomenal differences. It is what distinguishes solutions of nitric acid, whose identifying molecular composition is HNO_3, and hydrochloric acid, whose molecular composition is HCl, from water despite their striking phenomenal similarities. It is also what distinguishes distilled water from salty water, brackish water, and muddy water. Could a seventeenth century alchemist-chymist have discovered that salty water and distilled water have in common molecules of H_2O (and associated ions) or that *aqua regia* has a different molecular composition from water? The answer is clearly no. His scientific understanding of water is based upon its superficial sensible properties, and no amount of reflection on or investigation of these phenomenal properties with the equipment then available could reveal that water consists of two atoms of hydrogen chemically bonded to an atom of oxygen. Indeed, reflecting the continued dominance of the ancient Aristotelian theory of material substance, water was still regarded as an indivisible element as late as the mid-eighteenth century (Ball 1999; Levere 2001).[11] It was not until the late eighteenth century, when Antoine Lavoisier (1783) published his seminal paper "On the nature of water and on experiments that appear to prove that this substance is not properly speaking an element, but that it is susceptible of decomposition and recomposition" that the Aristotelian approach to chemical substance was finally abandoned.

Lavoisier's new theoretical framework for thinking about chemical substances in terms of elements and compounds laid the groundwork for modern molecular theory. Without it no one could have discovered that "pure" (distilled) water is composed of molecules of H_2O. Even so, it was not until the 1820s that the Swedish chemist Jöns Jakob Berzelius established that water consisted of two atoms of hydrogen combined with one atom of oxygen (Ball 1999); Dalton, the father of molecular theory, was convinced that it consisted of a single atom of hydrogen combined with a single atom of oxygen (HO).

---

[11] Aristotle believed that material substances were made up of four basic elements, earth, water, air, and fire. The first three of these elements correspond loosely to what we think of as phases (solid, liquid, and gas). From this perspective, when water evaporates it transmutes from water to a different chemical substance air, and when it freezes it transmutes from water to earth. European alchemists by and large accepted Aristotle's account of material substance, and their successors, seventeenth and eighteenth century chymists, incorporated aspects of it, especially the notion that water is a primary substance, into their work (Levere 2001; Newman 2006); in truth, there was not much of a distinction between alchemy and chymistry in the seventeenth century. The idea that the very same substance could take the form of a solid, gas, and liquid required a new theoretical framework (molecular theory) for thinking about material substance.

The identification of water with H_2O has enormous explanatory and predictive power. It explains what muddy water, brackish water, swamp water, distilled water, and even acidic solutions all have in common, despite their obvious phenomenal differences. It explains why liquids (such as *aqua fortis*) that resemble water in some ostensibly important ways, and indeed typically contain water molecules, are not water; they have the wrong molecular composition. In the context of modern chemistry, the identification of water with H_2O allows us to predict the behavior of water under very different physical and chemical conditions. It holds regardless of whether water is in its liquid, vapor, or solid phase, including its less familiar high pressure solid phases. No mere analysis of the observable properties of water could so definitively settle questions about the proper classification of such sensibly different substances as ice and steam, let alone such sensibly similar substances as nitric acid and distilled water.

Assertions such as 'water is H_2O', 'sound is a compression wave', and 'heat is a motion of molecules' are dubbed "theoretical identity statements" by philosophers. Theoretical identity statements provide the basis for scientifically compelling answers to 'what is' questions about natural kinds. But as the discussion above illustrates, they cannot be interpreted as ideal definitions even though they have the appropriate grammatical structure.[12] Instead of analyzing pre-existing human concepts, theoretical identity statements represent points at which our concepts changed as a result of empirical research in the context of fruitful theoretical frameworks for understanding natural phenomena. If they are not ideal definitions what are they?

During the first half of the twentieth century many philosophers of science classified them as stipulative definitions. Like ideal definitions, stipulative definitions supply necessary and sufficient conditions for membership in the class of items designated by a general term. Instead of dissecting pre-existing concepts, however, they represent explicit decisions to assign new concepts (meanings) to terms. The words may be familiar or newly coined for this purpose. Coining the term 'gene' for the basic unit of heredity provides a good illustration of assigning a new term to a theoretically important concept; Welhelm Johannsen invented the word 'gene' in 1909 to replace older terms (e.g., Mendelian "factor" or Darwinian "trait") that he thought were too theory laden (Sapp 2003, 135). The physicist's definition of 'work' as the product of the magnitude of an acting force and the displacement due to its action, on the other hand, provides a salient illustration of assigning an old familiar term to a theoretically important concept. In both cases the choice of term is arbitrary. Johannsen could have stuck with 'factor', and it

---

[12] As Chapter 3 explains, their grammatical form is a bit misleading because they cannot even be interpreted as genuine identity statements. But for purposes of this chapter, we can ignore this mismatch.

undoubtedly would have led to less confusion among students if physicists had selected a term other than 'work'. An important new theoretical concept needs a label and any term will do.

It is not clear, however, that identity statements such as 'water is H_2O' are stipulative in this sense. The decision to use the common noun 'water' for the molecular compound H_2O was not arbitrary. Armed with a promising new theoretical framework, nineteenth century chemists engaged in extensive empirical research for the purpose of *discovering* the molecular compositions of familiar chemical substances such as water. It is conceivable that the outcome of their investigations was that water is not H_2O; indeed, as mentioned earlier, Dalton was mistaken in concluding, on the basis of experiments, that water is HO. The question is why would someone interpret a theoretical identity statement such as 'water is H_2O' as a stipulative definition?

2.3.3 Diagnosing the Problem: A Defective Theory of Meaning and Reference

The view that natural kind terms can be defined is founded upon an old philosophical theory of meaning known as the descriptive theory. Commonly attributed to the seventeenth century philosopher John Locke, the descriptive theory dominated philosophical thought about the meaning and reference of general terms until the second half of the twentieth century and is still tacitly accepted by many non-philosophers, few of whom have even heard of it. It is important to appreciate why philosophers of language no longer accept this theory. For only then will it become clear why definitions provide excellent answers to 'what is' questions about non-natural (conventional and artificial) kinds but very inadequate answers to 'what is' questions about natural kinds.

According to the descriptive theory, the meanings of *all* general terms are fully exhausted by the concepts associated with them. On the classic version of the theory, concepts are identified with descriptions *qua* logical conjunctions of predicates supplying necessary and sufficient conditions for the application of a term. The concept associated with a general term thus fully determines its reference; anything satisfying the description falls into its extension and anything failing to satisfy the description does not fall into its extension. This explains the appeal of ideal definitions: They fully settle the question of whether something is a member of a natural kind.

On the downside, however, the descriptive theory rules out the possibility of discovering that one is wrong about the nature of the kind designated by a general term. Insofar as concepts (*qua* descriptions) fully determine whether something is a member of the category designated by a kind term, the only way the reference of the term can change is if the concept changes. If the concept (description) associated with a phenomenal kind term, such as 'water' or (presumably) 'life', changes in the right way the term's extension *ipso facto* changes too. The upshot is that nineteenth century chemists cannot be viewed as discovering that their predecessors (who believed that water is distinguished from other chemical substances not only as an indivisible element but also by its powers as a solvent) were wrong about phenomenal water. They merely changed the concept associated with the old, familiar term 'water' in a way that also changed the class of items to which it applies; nitric acid, along with a number of other chemical substances, ceased to be included in the extension of 'water'. This helps to explain why so many early twentieth century philosophers of science felt compelled to do violence to their intuitions and interpret theoretical identity statements as stipulative definitions analogous to the physicist's definition of 'work'.

From its inception, however, the descriptive theory faced problems. It is unable to distinguish natural kind terms from non-natural kind terms (such as 'bachelor', 'garbage', 'hammer', and 'American'). Unlike the former, the latter designate categories that are carved out by human conventions, interests, and concerns, and which would not exist had there been no human beings. Locke was fully aware of this difficulty and chose to bite the bullet (so to speak) and reject the distinction. In a revealing discussion (Locke 1689, Bk III, Ch. XI, Sec. 7), he argues that the seventeenth century debate over whether bats are birds has little scientific merit since the seventeenth century *concepts* of bat and bird are compatible with either position; on his view, the debate is merely verbal. In essence, Locke is demanding a stipulative definition of 'bat'. In hindsight, however, this seems misguided. The question of whether bats are birds is not just a matter of deciding which descriptions to associate with 'bat'. Scientists have discovered that the creatures called "bats" are more closely related to mammals than to birds. On the classic version of the descriptive theory, however, we must deny that this is the case.

Thomas Kuhn's infamous argument for the incommensurability of rival scientific theories (Kuhn 1962, Ch. IX), which undermines the idea that science makes progress, is founded on the classic version of the descriptive theory. As Kuhn points out, the descriptions associated with basic scientific terms undergo radical changes during scientific revolutions. In Newtonian dynamics, for instance, mass is conserved, whereas it is interconvertible with energy in Einstein's special theory of relativity. To the extent that these "definitions" – Kuhn's own word (p. 102) – are logically incompatible, 'mass' must be viewed as designating something different in Einstein's theory than in Newton's theory. Kuhn concludes that one cannot (as physicists typically do) claim that Newtonian dynamics is a special case of Einsteinian dynamics. Instead of extending Newton's theory to cases involving speeds close to that of light, Einstein has merely changed the subject (the class of items being talked about). But if (like most physicists) one views 'mass' realistically, as designating a theoretically fundamental, natural category, distinct from, for example, electric charge, then this conclusion seems wrong. Einstein taught us something that Newton did not know about the basic structure of physical reality.[13] Analogously, on Kuhn's account, nineteenth century chemists cannot be construed as discovering something new about the same old stuff that alchemists and chymists were talking about when they used the term 'water'; they merely changed the subject, in the process redefining 'water'. As with Locke's account of the controversy over whether bats are birds, these characterizations seem fundamentally mistaken. We do not think or speak about most scientific revolutions in this manner. We view them as representing defeasible discoveries, often about categories of items (water, bats, gold, etc.) to which human beings have historically referred and that are (in an important sense) delimited by nature as opposed to by human beings. However unlikely it might now seem, one cannot completely rule out the possibility of future chemists discovering that water is not H_2O.

The subject matter of a theory of meaning is language and thought. As a consequence, the success of a theory of meaning critically depends upon its capacity to explain indisputable facts about the way human beings use words and concepts. The failure of the classic version of the descriptive theory to account for the ways in which we speak and think about natural kinds is enough to seriously undermine it. But it is not catastrophic. Perhaps the differing ways in which we think and speak about natural kinds and non-natural kinds can be explained in a more holistic manner. Hilary Putnam established that this is not the case: He demonstrated that natural kind terms cannot be defined.

2.3.4 Why Natural Kinds Cannot Be "Defined"

Although it is beyond the scope of this chapter to explore Putnam's (1973, 1975) arguments against the descriptive theory in detail, it is important to appreciate their strength, particularly in view of the fact that a surprising number of philosophers of science still seem sympathetic to definitional approaches to explicating natural kinds. Part of the problem is that Putnam's arguments against the descriptive theory are sometimes misconstrued as supporting his alternative to it, namely, the causal theory of reference. While there are good reasons for doubting a purely

---

[13] On an instrumentalist account of theoretical kinds, Kuhn's argument goes through for 'mass' (which was treated as a theoretical term in Newtonian mechanics) but not for phenomenal kind terms such as 'water'; on this view, mass does not have a nature independently of what the theory says about it. The question of whether theoretical kinds should be given a realist or instrumentalist interpretation is unfortunately beyond the scope of this book; for an interesting discussion see Psillos (1999, Ch. 12).

causal theory of the reference of natural kind terms, there is little doubt that Putnam's argument against the classic version of the descriptive theory succeeds.

Readers lacking a background in philosophy will find the arguments in this section subtle and difficult. For those who still believe that life can be defined – that it is just a matter of being clever enough to come up with a definition which is counterexample proof – they are well worth slogging through. The heart of Putnam's case against the descriptive theory is captured in his famous Twin-Earth thought experiment. We are asked to imagine a fantastic planet (Twin-Earth) that is just like Earth except that the stuff that resembles water has a different chemical composition (abbreviated as XYZ). XYZ and H_2O have the same sensible properties, namely, being wet, transparent, odorless, tasteless, and a good solvent. The oceans, lakes, and rivers of Twin-Earth contain XYZ, as opposed to H_2O. Twin-Earth is occupied by people who are not only biologically, psychologically, and culturally like us but whose history also closely parallels our own.

The chemical compositions of the so-called water on Twin-Earth and the water on Earth are not discovered on either planet until the early nineteenth century. Let $Oscar_E$ be a typical speaker of English on Earth in 1750 and let $Oscar_T$ be a typical speaker of English on Twin-Earth. $Oscar_T$ and $Oscar_E$ not only look alike but also have the same backgrounds, experience the same feelings, and (most importantly) have the same mental (psychological and neurophysiological) states. Because the stuff known as water on both planets has the same sensible properties, Twin-Earthers and Earthlings have the same concept of water; they use the same identifying descriptions to pick out samples of water on their respective planets. Were $Oscar_E$ to visit $Oscar_T$, he would most likely believe that there is water on Twin-Earth. But as scientists on Earth and Twin-Earth later discover, in the nineteenth century, he would be wrong. The stuff that $Oscar_T$ calls "water" has the wrong chemical composition. It is not H_2O. It follows that the extension of the term 'water' is not fully determined by concepts in the mind. If it were, we would draw a different conclusion from the Twin-Earth example. We would conclude (on the basis of the sameness of their concepts) that $Oscar_E$ and $Oscar_T$ are talking about the same kind of stuff. Putnam has driven a wedge between the referents of natural kind terms and the concepts (*qua* identifying descriptions in our minds) that we associate with them. As he colorfully put it, "[c]ut the pie any way you like, 'meanings' just ain't in the head" (Putnam 1979, 227).

The Twin-Earth thought experiment shows that the extension of natural kind terms is not fixed by human concepts,[14] which undermines the rationale for

---

[14] This contrasts with the extension of *non-natural* kind terms such as 'bachelor' and 'garbage' whose meaning depends exclusively on human concepts. There is no empirical evidence that could undermine the claim that the two Oscars are talking about the same thing *if their concepts of bachelor are the same*. Being a bachelor is not a *natural* category; it depends upon human conventions, interests, and concerns. If there

definitional approaches to explicating natural kinds. If the classic descriptive theory were correct, we would have little choice but to interpret theoretical identity statements such as 'water is H_2O' as stipulative definitions; for it is clear that 'H_2O' does not represent an analysis of the ordinary, everyday concept of water and hence must represent a modification of the concept associated with 'water'. Commitment to the classic version of the descriptive theory explains why so many philosophers of science in the first half of the twentieth century, before Putnam's devastating critique of the classic descriptive theory for natural kind terms, felt compelled to interpret theoretical identity statements as stipulative definitions, despite the fact that they *seem* to represent *bona fide* scientific discoveries.

It is important to be clear about what Putnam's Twin-Earth thought experiment does and does not establish. It establishes that the concepts associated with at least some natural kind terms do not completely determine their extensions. Scientists are sometimes frustrated with the scientific implausibility of the Twin-Earth thought experiment: How could 'water' on Earth and 'twater' (as Putnam dubs it) on Twin-Earth differ in molecular composition and yet be, among other things, drinkable by humans on both planets? It is important to appreciate that such objections to Putnam's argument are irrelevant. Putnam is not making a scientific claim about water. He is making a claim about *language* and *concepts*. A theory of linguistic meaning should be neutral with respect to what our scientific theories tell us about the world; it should hold even if they are wrong. Language is used to describe many different kinds of situations, from factual, to hypothetical (e.g., what if Al Gore had been the US President in 2003), to fantastic (e.g., the adventures of the young wizard Harry Potter). It is the responsibility of a theory of meaning to account for the ways in which we think and speak about fantastic situations as well as hypothetical and factual situations. Putnam's Twin-Earth thought experiment uses a fantastic situation to demonstrate an important point about language: Concepts (*qua* descriptions in our heads) do not always fully determine the extensions of natural kind terms.

The Twin-Earth example has been extensively criticized. Although some, for instance, Mellor (1977) and Zemach (1976), challenge Putnam's intuition that XYZ is not water, it is significant that their counterintuitions are not widely shared. As John Dupré counsels (1993, 25) the most that can be said is that we are uncertain about what to say in such a "fantastic situation." But, and this is the crucial point, this uncertainty alone is enough to undermine a definitional approach

---

were no humans there would be no bachelors; in contrast, if there were no humans there would still be water and, presumably, life. In short, Putnam's argument against definition does not extend to non-natural kind terms, whose meanings are determined by human concepts.

to understanding natural kinds. If the classic theory were correct, we should not hesitate in drawing the conclusion that Mellor and Zemach urge upon us.

It is important not to construe Putnam's arguments *against* the descriptive theory as arguments *for* his alternative theory, the causal theory of reference. The Twin-Earth thought experiment shows that the reference of natural kind terms is not fully determined by human concepts. The latter leaves the problem of fleshing out what, other than concepts, could do the job. Putnam's solution is causation. The details of his version of the causal theory of reference are not important for our purposes. The basic idea is that the reference of a natural kind term, such as 'water', is fixed by complex causal relationships holding between a community of speakers and the natural world independently of whatever descriptions an individual speaker happens to associate with it (Putnam 1973, 1975). Someone who does not know that water is H_2O can thus succeed in referring to water. The ability to refer to water without understanding its nature opens up the possibility of making good sense of the idea that contemporary chemists are talking about the same old stuff that alchemists and chymists were talking about, even though their respective concepts of water are quite different.

Purely causal theories of reference, like Putnam's, face serious problems. Luckily alternatives to both the classic descriptive theory and the causal theory of reference are available. As discussed below, however, none of them is compatible with a definitional approach to explicating natural kinds.

The oldest alternative to the classic descriptive theory, which predates Putnam's work, is known as the cluster theory. Drawing its inspiration from Wittgenstein's famous analysis of the meaning of 'game', the cluster theory follows the classic theory in identifying concepts with descriptions but it analyzes them as loose clusters (versus logical conjunctions) of predicates. On the cluster theory, there are no sufficient conditions for belonging to a kind, and no single description can be taken as necessary. Being a member of the class of items designated by a natural kind term is just a matter of having enough (different versions tell different stories) of the pertinent properties. While not completely excluding the possibility that the concepts associated with some natural kind terms provide necessary and sufficient conditions for their applications, the cluster theory denies that this is typically the case, which is enough to frustrate the project of defining natural kind terms.[15]

In essence, the cluster theory is an early attempt to shore up the descriptive theory in the face of looming difficulties with the idea that general terms can be

---

[15] Like the classic version of the descriptive theory, the cluster theory still faces the problem of explicating the meaning and reference of natural kind terms solely in terms of concepts *qua* descriptions in our heads, and hence cannot escape the Twin-Earth thought experiment. Oscar$_E$'s and Oscar$_T$'s concepts of water are still the same in 1750; their concepts do not include the molecular composition of water (H_2O) and twater (XYZ), which were discovered in the nineteenth century.

defined. Like the classic version of the descriptive theory, however, the cluster theory still faces the problem of explicating the meaning and reference of natural kind terms solely in terms of concepts *qua* descriptions in our heads. As a consequence, it cannot escape Putnam's Twin-Earth thought experiment. Oscar$_E$'s and Oscar$_T$'s eighteenth century concepts of water do not include the molecular compositions of water and twater, which was not discovered until the nineteenth century. Accordingly, the cluster of properties (descriptions) that they associate with water is the same. Viewed in this light, Putnam's thought experiment shows that neither version of the descriptive theory provides a satisfactory account of the meaning and reference of natural kind terms.[16] It is thus not surprising that the descriptive theory has fallen out of favor among philosophers of language.

Mixed causal-descriptive theories dominate current philosophical thought about the meaning and reference of natural kind terms. These hybrid theories reject the extremes represented by descriptive theories and (purely) causal theories of reference (such as Putnam's). The reference of natural kind terms is determined through the interplay (different versions tell different stories) of description and causation; concepts in our heads coupled with (mind-independent) causal relations between language and the world *jointly* determine reference. A salient illustration is Chalmers and Jackson's (2001) (aptly named) two-dimensional semantics, which explicitly rejects the identification of concepts with descriptions and moreover discusses the untoward consequences of this rejection for defining natural kind terms (pp. 320–323); as they note, identifying the extension of a natural kind term requires "sufficient empirical information about the actual world" (p. 323). In short, none of the alternatives to the classic descriptive theory supports a traditional definitional approach to explicating natural kinds.[17] The scientific project of defining life is founded upon a failed theory of the meaning and reference of natural kind terms and hence ought to be abandoned.

2.3.5 Is Life an Exception to the Rule?

Putnam's Twin-Earth argument leaves open the possibility that the *concepts* associated with some natural kind terms provide necessary and sufficient conditions for their application; his argument defeats the descriptive theory insofar as it establishes that this is not *in general* the case. So it seems that one cannot preclude the possibility that 'life' is a rare natural kind term whose extension (like that of

---

[16] The search strategy that I develop in Chapter 8 – for searching for potentially biological anomalies using tentative (versus defining) criteria for life – bears an uncanny resemblance to the cluster theory. There are critical differences, however.

[17] For a more detailed discussion of contemporary alternatives to the classic descriptive theory of the meaning and reference of natural kind terms, including Richard Boyd's (e.g., 1999) interesting homeostatic property cluster theory, see Cleland (2012).

non-natural kind terms) is fully determined by concepts after all. There are compelling empirical reasons for thinking that this purely hypothetical possibility is very unlikely.

Our concept of life has changed in important and sometimes radical ways through-out history (Lange 1995). For a long time, self-movement was thought to be a defining characteristic of life. From ancient times until as late as the seventeenth century, a debate simmered over whether stars and other celestial bodies were alive on the grounds that they are self-moved and their motion resembles that of birds or fish. Moreover, until the mid-nineteenth century it was not clear that fungi should be classified as living since they do not display what were then taken to be basic features of life, namely, movement, growth, and reproduction. Indeed, the "fungus-stone," which resembles a large gray stone and is so tough that a saw is required to cut it, led many early scientists to classify fungi as minerals (Ainsworth 1976, 32). The nineteenth century discovery (by Pasteur and others) that infectious diseases are caused by organisms (bacteria) that are invisible to the naked eye represents yet another profound change in our concept of life; in the Middle Ages infectious diseases were attributed to such things as bad air, supernatural influences, and humoral imbalances. The nineteenth century also witnessed the formulation of Darwin's theory of evolution, with its emphasis on the interrelatedness of all life on Earth; in contrast, earlier views treated species as fixed and eternal. With the advent of molecular biology in the late twentieth century, our concept of life changed yet again, becoming closely coupled to the molecular composition and architecture of organ-isms – to the complex cooperative arrangement between proteins and nucleic acids. It is hardly a coincidence that NASA scientists currently favor the chemical Darwinian definition (Joyce 1994). To put the situation into stark perspective, the idea of a seventeenth century biologist coming up with the chemical Darwinian definition is no more plausible than the idea of a seventeenth century chemist coming up with a definition of water as H_2O.

Unlike the present-day chemical concept of water, however, the biological concept of life will almost certainly continue to change in unanticipated ways. For modern biology does not have a truly general theory of life analogous to the chemist's molecular theory of chemical substance. In the context of modern chemistry, the unique association of water with the molecular compound H_2O allows scientists to explain and predict properties of water here on Earth and, most importantly, anywhere else in the universe. These properties include being a good solvent, expanding when frozen, and remaining a liquid over a wide range of temperatures. Many of these properties were thought to be distinct-ive of water before the advent of molecular theory, and the ability of molecular theory, coupled with considerations from chemical thermodynamics (e.g., Gibbs's law), to explain them played a significant role in the acceptance of the claim that water is H_2O (Needham 2002). The association of water with H_2O also allows scientists to explain and predict the behavior of water under chemical and physical conditions (very high pressures and temperatures) that do not naturally occur on Earth. The empirical fruitfulness of the claim that water is H_2O helps to explain why contemporary scientists believe that they have a good understanding of the universal nature of water, and hence a compelling scientific answer to the question 'what is water?'

But as ongoing disputes over the "definition" of life underscore, scientists are not even in agreement about something as basic as whether life is (like water) a physicochemical kind, as opposed to an abstract functional kind. As Chapter 5 discusses, the inability to decide this issue is in large part due to the limited nature of our experience of life. There is overwhelming empirical evidence that all known life on Earth today descends from a last universal common ancestor (LUCA), which means that we are dealing with a single example of life. One cannot safely generalize from a single example of life to all of life, wherever and whenever it may be found in the universe. For under such circumstances there is little empirical basis for differentiating characteristics common to all of life from those deriving from physical and chemical contingencies at the time of the origin and early development of our particular form of life on the ancient Earth. The situation is exacerbated insofar as many scientists still tacitly privilege complex multicellular eukarya (especially plants and animals) in their theorizing about life even though, as Chapter 5 explains, they are now recognized to be an unrepresentative variety of Earth life. In contrast, there are no empirical reasons for suspecting that the water found on Earth is unrepresentative of water considered generally, which makes theorizing about water on the basis of samples from Earth much less problematic (although still, of course, subject to Putnam's hypothetical Twin-Earth argument). The point is there are *empirical* reasons for suspecting that none of our *contemporary* scientific concepts of life contains identifying knowledge of the kind. As a consequence, even supposing for the sake of argument that some natural kind terms can, like non-natural kind terms, be successfully defined, there are compelling scientific reasons for thinking that 'life' is not among them.

2.4 Concluding Thoughts

This chapter began by critically evaluating popular definitions of life and ended by concluding that the viability of the scientific project of defining life rests upon a defective theory of the meaning and reference of natural kind terms. More specifically, the most informative ("ideal") definitions analyze concepts that we already possess, and hence are intrinsically incapable of revealing the objective underlying nature (or lack thereof) of categories designated by natural kind terms. The upshot is that if (as seems likely, but not certain) life is a natural kind, scientific attempts to define 'life' are self-defeating. For when a scientist seeks an answer to the query 'what is life?' she is not interested in an analysis of the contemporary human concept of life. She wants to know what life *truly* is: What bacteria, slime molds, fungi, fish, trees, birds, and elephants all have in common that distinguishes them from nonliving physical systems, and what any system, however different, must have in common with these organisms in order to qualify as life. Analysis of the twenty-first century concept of life is unlikely to be of much help in this endeavor. Indeed, if our current concept of life is badly mistaken, which (as Chapter 5 argues) is a very real possibility, a definition will serve only to entrench our misconceptions, making it more difficult to jettison them in light of new empirical discoveries. A definition of life is thus likely to blind us to truly novel forms of life should we be so fortunate to encounter them either here on Earth – as a heretofore unrecognized "shadow biosphere" – or elsewhere in the solar system. We will return to these and related issues in Chapters 6, 7, 8, and 9.

I have argued that a definition of life cannot provide a scientifically compelling answer to the question 'what is life?' The alternative is a scientific theory. But what exactly is a scientific theory? The next chapter discusses the structure and function of successful scientific theories. Many scientists and some philosophers (who are aware of the difficulties facing definitions of natural kind terms) believe that there is a close logical relationship between scientific theories and definitions. They defend what amount to nonstandard definitions of life, that is, definition-like statements supplying necessary and sufficient conditions for life whose authority derives from empirical investigations, as opposed to analyses of concepts. As Chapter 3 argues, however, there are important structural differences between definitions and scientific theories. One can no more compress a scientific theory of life into a nonstandard definition of life than one can compress it into a traditional definition of life. Unlike definitions of any sort, scientific theories do not fully determine the membership of the natural kinds that they subsume, and their inability to do this is part of what makes them such powerful tools for exploring nature. Chapter 3 concludes that the definitional project, in all its nefarious guises, needs to be abandoned.

3

What Is a Scientific Theory?

3.1 Overview

Not everyone who advances a so-called definition of life has in mind the traditional notion of definition. This is especially true of scientists. Biochemist Steve Benner (2010), for instance, contends that definitions encapsulate theories, and speaks of the need for formulating a "definition-theory of life" (p. 1022). But it is also true of some philosophers who are well aware of the limitations of traditional definitions. As an illustration, Mark Bedau (1998) presents a "definition" (his term) for life and characterizes it as the "general form of my theory of life" (p. 128). Definitions of this sort are *nonstandard* in the sense that their authority does not derive from analysis of human concepts or alternatively mere stipulations of meaning. Their acceptability depends upon successful empirical investigations. Nonstandard definitions nonetheless resemble traditional definitions *structurally* insofar as they supply necessary and sufficient conditions (identifying descriptions) for membership in a presumed natural kind. Some recent proposals for "defining" life, briefly discussed in Section 3.5, are even more radical, rejecting the received view that a central function of definition is classification; on such a proposal, a definition of life need not even provide necessary and sufficient conditions for life.

The central goal of this chapter is to explore the composition, structure, and function of scientific theories for the purpose of (1) putting the final nail in the coffin of definitional approaches to understanding life and (2) providing important background information for understanding how successful scientific theories develop, which is the topic of Chapter 4. The oldest and most influential philosophical characterizations of the composition and structure of scientific theories are the *syntactic conception* (Section 3.2) and the *semantic conception* (Section 3.3).[1] These conceptions hold forth the greatest promise of establishing

---

[1] Nowadays these characterizations of scientific theory are often called "views." The reason for the odd language is that there is a gap between actual scientific theories and the way they are, dare I say it,

a close logical relationship between nonstandard definitions and scientific theories, which is why this chapter focuses on them.[2] The syntactic and semantic approaches assume that there is no significant structural difference between a scientific theory and a mathematical theory. They reconstruct scientific theories along the same lines as those of mathematics as either deductively closed axiom systems (the syntactic view) or set theoretic structures (the semantic view).

Section 3.4 compares and contrasts nonstandard definitions with scientific theories reconstructed in accordance with the syntactic and semantic conceptions for the purpose of pin pointing logical differences between scientific theories and nonstandard definitions. As will become clear, nonstandard definitions cannot be logically inferred from scientific theories, and scientific theories cannot be encapsulated in nonstandard definitions. As Section 3.4 also explains, there are equally important differences in the functions served by definitions and scientific theories. The primary function of definition (traditional and nonstandard) is classification. In contrast, classification is neither the only nor the most important function served by scientific theories; the most obvious functions of scientific theories are prediction and explanation, but they are not the only ones. Most intuitively compelling classification schemes are incapable of supporting the functions of scientific theories. Moreover, as will become clear, the classifications provided by scientific theories are not like those of definitions. Definitions fully determine (via necessary and sufficient conditions) the extension of natural kind terms whereas the classifications of scientific theories are open ended and fluid. This flexibility is not a defect. It is what allows scientific theories to accommodate novel and sometimes startling empirical discoveries. Indeed, as we shall see, pursuit of necessary and sufficient conditions for partitioning a physical system into life and nonlife is more likely to hinder than facilitate the development of a truly general theory of life.

conceptualized by philosophers of science. This helps to explain why I prefer the older term 'conceptualize' to 'view'. Basically, the older term more clearly indicates that both characterizations of scientific theory are idealizations extracted from real scientific theories for purposes of laying bare an underlying structure for logicomathematical analysis; without the latter, the claim that there is a close structural relationship between scientific theories and nonstandard definitions, which have a very precise structure, cannot even be evaluated, let alone justified.

[2] On the newer pragmatic accounts, the structure of scientific theories is treated as highly complex, variable, and amorphous. As a result, they cannot support the claim that there is a close structural relationship between scientific theories and definition-like statements supplying necessary and sufficient conditions for natural kinds. For this reason, I do not discuss them in this chapter.

3.2 The Syntactic Conception of Scientific Theories

The earliest philosophical analysis of scientific theories is known as the syntactic conception. The roots of the syntactic conception lie in the work of the logical positivists,[3] with the penultimate version, dubbed the "received view" (Putnam 1962), formulated in the middle of the twentieth century by Rudolf Carnap and Carl Hempel; see Suppe (1977, Chs. I and II) for a detailed discussion of the history of its development. Advocates of the syntactic approach analyzed the structure of scientific theories along the lines of classical formulations of Euclidean geometry as deductively closed axiom systems. Deductively closed axiom systems consist of collections of sentences having a special logical structure: Every sentence in the system is deducible from a small number of basic axioms or postulates whose truth is assumed. The basic components of a deductively closed axiom system are (i) a small number of primitive (undefined) nonlogical terms (e.g., 'point' and 'line'), by means of which other useful nonlogical terms (e.g., 'straight line', 'circle') can be defined, and (ii) a small number of postulates or axioms (e.g., a straight line can be drawn from any point to any point) that are assumed to be true; the latter cannot be proven within the system. An indefinitely large number of additional sentences, such as the Pythagorean theorem, can be logically derived from these components. The axioms and undefined terms thus serve to simplify and systematize a much larger collection of geometrical claims.

Advocates of the syntactic conception believed that reconstructing scientific theories along the lines of those of mathematics would reveal how successful scientific theories simplify and systematize domains of *phenomenal* (perceptual) experience for purposes of description, prediction, and explanation. Their initial efforts focused on highly successful foundational theories of physics, which are especially notable for their mathematically precise laws. Newton's theory of terrestrial motion provides a salient illustration. Glossing over details, Newton's three laws of motion are treated as basic postulates. Mass and position in space and time are taken as basic concepts. Force, velocity, and acceleration are nonbasic because they can be "defined" in terms of the former; Newton's second law of motion (force = mass × acceleration) defines force in terms of mass and acceleration, the latter of which is defined as changing rate of change in length traveled per unit of time. Defenders of the syntactic conception conjectured that any mature scientific theory could be reconstructed (at least in principle) as a deductively closed set of postulates formulated in first order logic (Suppe 1977).

---

[3] The logical positivists were a group of philosophers, mathematicians, and scientists who met together in Vienna, Austria in the 1920s–1930s to discuss the nature of science and human knowledge; they were known as the "Vienna Circle."

On the syntactic conception, the main difference between a scientific theory and a mathematical theory is the need to connect the former to real world phenomena through observation. This was achieved by dividing the terms of a reconstructed scientific theory into two classes, those that are directly "definable" in terms of observation (*observational terms*) and those that are not (*theoretical terms*). The meanings of observational terms are given in terms of human sensory experience.[4] Providing an interpretation for theoretical terms was more challenging. For the classes of entities (theoretical kinds) designated by basic theoretical terms (e.g., 'mass' in Newtonian mechanics, and 'gene' in genetics) are remote from human sensory experience and yet critical to a theory's capacity to simplify and systematize the domain of natural phenomena to which it applies. As illustrated by Newton's second law of motion, basic theoretical terms designate the theoretical kinds characterized and interrelated by the fundamental laws (postulates) of the theory.

Early versions of the syntactic conception attempted to fully define theoretical terms by means of *correspondence rules* connecting them directly to logical complexes of observational terms. Correspondence rules were typically interpreted as operational definitions (Bridgman 1938); see Section 2.3.1 for more on operational definitions. They were supposed to supply separate definitions for each basic theoretical term in terms of observation. A correspondence rule for Newton's theoretical concept of mass could involve weighing an object on a scale since one can calculate the mass of an object from its weight using Newton's theory. Correspondence rules were never intended as "defining" (phenomenal) natural kinds, such as weight or water in terms of theoretical kinds;[5] correspondence rules are supposed to explicate theoretical kinds in terms of observables, *not vice versa*. The upshot is that theoretical identity statements (see Section 2.3.2), such as 'weight is the force on an object due to gravity' and 'water is H_2O', cannot, on the syntactic conception, be interpreted as providing nonstandard definitions for natural kinds (weight and water) via correspondence rules, which of course does not mean that theoretical identity statements could not be interpreted as nonstandard definitions of some other sort. We will return to this issue shortly.

---

[4] There was disagreement over the objects of "sensory experience." Some identified them with sense data (e.g., patches of color) or complexes of sense data. Others (so-called physicalists) took them to be observable characteristics of familiar material things.

[5] Theoretical kinds are natural kinds *if* they reflect genuine divisions in nature. Theoretical kinds are not directly perceptible, however, which raises the question of whether they are merely convenient theoretical instruments (heuristic devices) for organizing human experience, as opposed to authentic natural categories that would exist even if there had been no humans to postulate them. This issue is especially thorny because theoretical kinds (e.g., phlogiston and the aether) are sometimes eliminated during theory change. For purposes of side-stepping this issue, which is not relevant to issues discussed in this book, I will restrict the term 'natural kind' to phenomenal (perceptible) kinds.

On the syntactic conception, the empirical interpretation of a scientific theory, supplied by correspondence rules, is an integral part of the theory itself. The correspondence rules turn an abstract logical system into a scientific theory, that is, a theory about real world phenomena. Problems arose early on. Correspondence rules were supposed to fully define theoretical terms operationally. The upshot is that modifying a correspondence rule for a theoretical term, such as 'mass', in light of the development of alternative methods or instruments (e.g., balance scale, hydraulic scale, and spring scale) for measuring what it designates, produces a different theory; one has literally changed the definiens, and hence the definition (which supplies a *specific* set of operationally necessary and sufficient conditions) for the physical application of the theoretical term.

The difficulties involved in giving empirical meaning to theoretical terms were not the only challenge faced by the syntactic conception. Different axiomatic formulations of the same theory – for example, the Hamiltonian formulation of Newtonian mechanics, which takes momentum, as opposed to mass, as a basic theoretical concept – count as distinct theories on the syntactic conception; they are built from different basic concepts and hence cannot be reconstructed as the same axiomatic system. Successive revisions to accommodate these and other difficulties culminated in the penultimate version of the syntactic view in which theoretical terms are *implicitly* defined by their role in the fundamental laws of the theory and given only *partial* empirical interpretations by correspondence rules connecting them to observational terms. To cut a long story short, advocates of the syntactic approach were reluctantly forced to concede that the meanings of theoretical terms could not be packed into definitions supplying necessary and sufficient conditions for their application after all. The inability to fully define theoretical terms in terms of observables was disappointing but not fatal to their analysis of scientific theories.

The syntactic conception dominated philosophical thought about scientific theories through the middle of the twentieth century. By reconstructing scientific theories as formal axiomatic systems, philosophers hoped to reveal how scientific theories are able to successfully simplify and systematize domains of natural phenomena. Just as metamathematical analysis of an axiomatized mathematical theory, such as Euclidean geometry, may reveal defects in the theory (axiomatic incompleteness, undefinable nonbasic terms, etc.) so supporters of the syntactic approach believed that axiomatic reconstruction of a scientific theory could reveal defects and suggest ways in which it might be improved to optimize descriptive, predictive, and explanatory success. Formalizing scientific theories as deductively closed axiom systems also held forth the promise of clarifying logical relations of identity and subordination among theories, such as whether Newton's theory of motion is a special case of Einstein's special theory of relativity.

Attempts were made to axiomatize a variety of scientific theories from different disciplines, but with limited success. The most promising (but, alas, typically incomplete) axiomatizations were of theories in physics, especially Newtonian mechanics (see, e.g., Hermes 1938, 1959; Simon 1970), whose foundations had been extensively studied. Another famous example from physics is von Neumann's axiomatization of quantum mechanics, which showed that wave and matrix mechanics were the same theory (von Neumann 1932). Theories of the special sciences[6] proved especially recalcitrant, however; see Williams (1970) for the best-known attempt at formalizing Darwin's theory of evolution. These failures, in conjunction with the unresolved difficulties mentioned earlier, strongly suggested that scientific theories could not be satisfactorily reconstructed as deductively closed systems of fundamental laws after all. By the 1970s, many philosophers had abandoned the syntactic conception and were seeking an analysis more closely conforming to the practicing scientist's conception of theory.

3.3 The Semantic Conception of Scientific Theories

Scientists rarely present theories as deductively closed axiom systems. On the other hand, they commonly articulate them through models. Some well-known examples are the billiard-ball model of a gas, the Bohr model of the atom, the Lotka–Volterra model of predator–prey relations, the double helix model of DNA, and the *lac operon* model of gene regulation. It is thus hardly surprising that many mid-twentieth century philosophers concluded that models (versus deductively closed collections of formalized fundamental laws) are the basic units for understanding scientific theorizing.[7] The philosophical question is what (if anything) do the diverse models used by scientists have in common and how do they simplify and systematize natural phenomena for purposes of description, prediction, and explanation?[8] In search of insight into these questions, philosophers of science turned to the newly emerging field of metamathematics, which uses the methods of mathematics to study mathematical theories.[9] The following material is difficult for anyone lacking a background in set theory and mathematical logic. But it is well worth the effort. For the semantic conception of scientific theories, which remains highly influential among philosophers, provides no support whatsoever for the

---

[6] Technically speaking, the term 'special sciences' includes all of science except for fundamental physics. It is most often used, however, to refer to biology and the social sciences.

[7] Not all philosophers of science view models as embodying theories. Morgan and Morrison (1999), for example, argue that modeling is primarily a tool for solving scientific problems and is only loosely dependent on theory. It is probably true to say that some scientific models serve as representations of scientific theories and others do not. Because our concern in this chapter is with model-based conceptions of scientific theory I will ignore this issue.

[8] For an excellent discussion of models and their diverse roles in scientific practice see Weisberg (2013).

[9] David Hilbert and Kurt Gödel are the best-known founders of metamathematics.

claim that scientific theories can be "captured" in nonstandard definitions for the natural kinds that they explicate.

Originating in the work of Suppes (1967), Van Fraassen (1970), Sneed (1971), and Suppe (1977, 221–239), the semantic conception is the oldest and most popular of the model-based approaches to understanding scientific theories. Under the semantic conception, the informal models actually used by scientists are rationally reconstructed as logicomathematical objects (models in the formal sense of mathematical logic). Underlying the semantic conception is the view that the relevant concept of model for the sciences is the same or very similar to that found in mathematics. The best-known versions of the semantic conception reconstruct scientific models as either set theoretic structures or state space structures; see Suppe (1989, Ch. 1) for a discussion of different formulations of the semantic conception. For the sake of simplicity, I will confine my discussion to the set theoretic version; for our purposes, nothing of consequence follows from doing so.

A set theoretic structure, $S = \langle U, O, R \rangle$, is a highly abstract, composite object consisting of (i) a nonempty set U of items (the domain of S), (ii) a set O of operations on U (which may be empty), and (iii) a nonempty set R of relations on U; see Machover (1996, Chs. 8 and 10) for more detail on the metamathematical concept of a set theoretic structure. The transition from real world phenomena to scientific models to the set theoretic structures of the semantic conception involves several stages of abstraction.[10] As an overly simplified illustration, consider the billiard-ball model of the kinetic theory of gases. Real gases have many properties, some of which vary depending upon the specific gas under consideration and others of which are universal but nevertheless not thought to be theoretically relevant to reasoning about gases. The billiard-ball model simplifies real gases by abstracting a small subset of characteristics, namely, shape, size, and elasticity, thought to be especially salient for purposes of reasoning scientifically about real gases. This is not the end of the process of abstraction, however. The properties privileged by the model are idealized, that is, conceived as uniform in size and shape, as well as being perfectly elastic, even though real gas molecules vary in all these properties.[11]

For advocates of the semantic conception this is not the end of the process of abstraction. Highly idealized, billiard-ball-like entities are still too concrete. The semantic conception requires structures more closely approximating the purely formal, set theoretic objects of metamathematics. At this rarefied level of

---

[10] Some philosophers (e.g., Weisberg 2007, 2013) distinguish abstraction (*qua* simplification) from idealization (as distortion). To avoid getting bogged down in this debate, I treat all three as modes of abstraction. Doing so simplifies discussion and does not affect the arguments in this book.

[11] Simplification and idealization occurs in the other direction too. Real billiard balls have many properties other than size, shape, and elasticity, for example, being made of ivory and having numbers on them. Moreover, real billiard balls are not perfectly elastic and vary (albeit modestly) in size and shape.

abstraction, the connection between the phenomena being modeled (real gases) and the model all but disappears. What matters is that the model and the modeled share the same abstract pattern of relations – the same set theoretical structure. In the words of Redhead (2001), "it is this abstract structure associated with physical reality that science aims . . ." (p. 75). Many defenders of the semantic conception, however, concede that *content* (what the theory says about the world) as well as structure matters to scientific reasoning about the world, but they are not very specific about where to stop the process of abstraction.[12] As we shall see, this vagueness generates problematic obscurities.

How are the rarified models of the semantic conception related to familiar linguistic-mathematical formulations of scientific theories? A model is character-ized as a set theoretic structure in which all the sentences of the theory – basic and derived principles, which may consist of mathematical generalizations, e.g., $F = ma$, or be more qualitative, e.g., law of conservation of mass (which says that the quantity of matter remains constant in an ordinary chemical reaction) – turn out true *under an interpretation*; following the mathematical logician Alfred Tarski (e.g., 1944, 1951), the model is said to "satisfy" the theory. An *interpretation*, in the logician's sense, is a function mapping the basic logicolinguistic components of a scientific theory (e.g., *n*-ary *predicates*) to the basic set theoretic elements of a structure (*n*-ary set theoretic *relations*).[13] The semantic conception thus departs from the syntactic conception in no longer incorporating a unique interpretation (via correspondence rules) within the body of a scientific theory. The models that satisfy the theory may do so by means of different interpretative functions. Furthermore, under different interpretations, the very same model may satisfy different theories. Because many different models may satisfy the same linguistic formulation of a scientific theory, the semantic conception individuates scientific theories in terms of unique classes (known as "families") of models. All the different representational possibilities for modeling a given theory are included

[12] Suppe (1989), for instance, restricts the process of abstraction to the "causally possible" (p. 84), which is a subset of the logically possible of the mathematician. But the concept of causal possibility involved is left obscure; neo-Humeans conceive of the causal relation as involving little more than *de facto* regularity whereas realists about causation construe it in terms of much stronger, counterfactual supporting, relations (the accounts differ). There is also the question of whether one can analyze the content of a scientific theory in terms of a general, one-size-fits-all, concept of causal relation; causation in quantum mechanics (assuming that one can make good sense of it) is different from causation in classical physics insofar as it is intrinsically indeterministic and (in quantum entanglement) involves the direct interaction of objects widely separated in space without an intervening causal chain; the latter is known as nonlocality.

[13] See Suppes (1957) and McKinsey et al. (1953) for a formalization of Newtonian mechanics under the semantic conception; the reader is warned that the result does not look much like Newton's original formulation of the theory, although he/they argue that it is equivalent. Part of the problem is that Newton did not formulate his theory as a model. As anyone who has skimmed Newton's *Principia* knows, he based his formulation closely on Euclidean geometry, which explains why it is easier to reconstruct Newtonian mechanics under the syntactic conception. But as we have seen, the syntactic conception is afflicted with a number of very serious problems, and moreover, contemporary science has moved away from axiomatic formulations of scientific theories.

in the family in virtue of sharing the same abstract set theoretic structure. To the extent that this higher order, set theoretic structure is viewed as a logicomathematical object, representational features that are not purely (formal in the scientist's version of the theory) are treated as primitive, which means that what the theory says about the world doesn't matter; if it did, not every model in the family of models would qualify as a version of the scientific theory concerned.

Each model in the family of models corresponding to a scientific theory satisfies a *theoretical definition* consisting of a logical conjunction of the basic laws of the theory. Early versions of the semantic conception followed the syntactic approach in requiring the fundamental laws of a scientific theory to be translated into a formal language, and distinguished theories (*qua* logicolinguistic entities) from models (*qua* logicomathematical structures). Later versions (e.g., Giere 1979; Van Fraassen 1980, 1989) loosened this restriction, permitting the basic laws of a theory to be expressed informally in natural language. (As will be discussed in the following section, this is what contemporary philosophers usually have in mind when they claim that scientific theories can be compressed into definitions.) Nevertheless, to the extent that the semantic conception embraces a model theoretic analysis of scientific theory (in the strict sense of mathematical logic), what matters is the logical structure of the theory not how it is characterized linguistically. Indeed, one could in principle by-pass the linguistic formulation of a theory and specify the corresponding class of models directly via the formal machinery of set theory. Viewed from this perspective, linguistic formulations of theories represent little more than heuristic devices. Yet, as discussed in Section 3.4 and emphasized in the next chapter, the key to the success of a scientific theory is not its set theoretic structure but its content – what it actually says about the material world. The latter is critical to, among other things, its explanatory and predictive capacities.

As might be expected, model theoretic analyses of mathematical theories provide especially clear analogs of how the semantic approach is supposed to work for scientific theories. The "standard model" for Peano Arithmetic (a highly intuitive, traditional axiomatization of the natural number system) consists of the set of natural numbers ($\mathbb{N}$) and includes two operations, addition (+) and multiplication ($*$), and two relations, equal-to (=) and less-than ($<$) [14] Any structure isomorphic to this structure comprises a *bona fide* model of the natural number system. An isomorphism (in the technical mathematical sense) is a mapping or function from the elements of one structure to those of another that is (i) one-to-one and onto and (ii) preserves the operations and relations from the first structure to the second. The mathematical concept of isomorphism thus fleshes out the intuitive

[14] This is a bit of a simplification, but it does not matter for our purposes in this chapter.

concept of identity of structure. In contemporary mathematical logic, the natural number system is identified with the family of all structures isomorphic to the "standard model." The standard (Peano) model is distinguished from the others only heuristically, as the intuitively compelling prototype. It is easy to design nonstandard models of arithmetic; an abacus, which represents numbers as beads strung on parallel wires and arithmetic operations as manipulations of beads, provides a good illustration of a physical model of arithmetic.

Unlike mathematical theories, however, the success of a scientific theory depends upon its connection to observable phenomena. On the syntactic conception, the connection between theory and observation is accomplished through correspondence rules. The semantic approach accomplishes this through *theoretical hypotheses* postulating either logical relations of isomorphism (Suppes 2002; Van Fraassen 1980), or homomorphism (Lloyd 1988), or partial isomorphism (Da Costa and French 2003), or informal relations of similarity (Giere 1988; Teller 2001) between *replicas* (models abstracted directly from a targeted phenomenal system) and the models in the class defined by the theoretical definition. In essence, the replica is hypothesized to satisfy the theoretical definition and hence be a member of the class of models individuating the theory. If the replica satisfies the theoretical definition then the target phenomenal system is said to fall within the scope of the theory. The idealized replica thus replaces the target system for purposes of categorization, prediction, and explanation. Characteristics of the target system that cannot be further decomposed into structural relations among basic components of the replica are treated as primitive or discarded as irrelevant; all that matters is whether the primitive elements of the replica can be mapped in the pertinent way (e.g., one-to-one and onto) to the elements of the standard model for the family of models individuating the theory. There is thus the danger that the replica does not provide an adequate representation of the target system after all – that characteristics critical to reasoning scientifically about the phenomena concerned have been left out of the replica; see Frigg (2006) for more on this issue.

It is telling that the switch from rationally reconstructing informal scientific theories as deductively closed axiom systems to reconstructing them in terms of logicomathematical structures parallels the switch in mathematical logic from reconstructing mathematical theories proof theoretically to reconstructing them model theoretically. Philosophers have had a love affair with mathematical logic since the beginning of the twentieth century. Fondness for mathematical logic, however, does not fully explain why philosophers abandoned the syntactic conception for the semantic conception. Switching to a model theoretic conception circumvented the seemingly intractable problem of providing a unique empirical interpretation for each theoretical term of a theory via a correspondence rule. On the semantic conception, the sentences of a scientific theory are interpreted holistically, as descriptions of complex structures; the basic terms of the theory cannot be separately defined in terms of observation. Furthermore, as logicomathematical objects, structures can be studied independently of a given linguistic formulation of a theory. They can also be characterized linguistically in different ways. Thus, the semantic conception is able to overcome another serious difficulty with the syntactic conception, namely, the ease with which one can formulate alternative axiomatizations of scientific theories. So long as two formulations of a theory (e.g., the Hamiltonian and classical formulations of Newtonian mechanics) pick out the same family of set theoretic models, they can be construed as expressing the same theory; for on the semantic conception a scientific theory is identified with the family of models. Unlike concrete physical systems, models *qua* logicomathematical structures are minimal, stripped (so to speak) to the bone. The underlying and highly problematic assumption, of course, is that the theoretical and empirical successes of scientific theories depend almost exclusively upon formal characteristics of the material world: In the words of Ladyman and Ross (2007), "There are no things. Structure is all there is" (p. 130).

3.4 Scientific Theories and Definitions

The reader will recall that nonstandard definitions of life have the logical structure of "ideal definitions" (Section 2.3). They determine the extension of the term 'life' by supplying necessary and sufficient conditions (identifying descriptions) for being a living thing. Nonstandard definitions differ from ideal definitions in an important way, however. Their authority is allegedly founded upon scientific discoveries, as opposed to human concepts and word meanings. This helps to explain the appeal of nonstandard (or as they are sometimes dubbed "scientific") definitions of life to astrobiologists. A nonstandard definition of life holds forth the promise of settling the issue of whether a puzzling geochemical system found on Mars or Titan is a genuine case of life on *scientific grounds*.

The most promising candidates for nonstandard definitions are *theoretical identity statements* and the *theoretical definitions* of the semantic conception of scientific theories. As Section 3.4.1 discusses, nonstandard definitions cannot be interpreted as theoretical identity statements. Among other things, the expression 'theoretical identity statement' is a misnomer. Theoretical identity statements do not assert genuine identities, and hence do not supply necessary and sufficient conditions for being a member of a natural kind, such as water and (presumably) life. The question of whether a nonstandard definition for a natural kind could encapsulate a scientific theory encompassing the natural kind concerned is tackled in Section 3.4.2. As will become clear, the answer is "no." While the theoretical definitions of the semantic conception encapsulate entire scientific theories, they do not provide necessary and sufficient conditions for any of the natural kinds that scientific theories subsume. As a consequence, a nonstandard definition *of life* cannot be viewed as supplying a tentative, let alone a mature, scientific theory of life.

3.4.1 Scientific Theories Do Not "Define" Natural Kinds

At first glance, theoretical identity statements, such as 'water is H_2O' and 'sound is a compression wave', seem to be promising candidates for nonstandard definitions (of, respectively, water and sound). They are commonly construed as asserting contingent identities (between phenomenal natural kinds and the theoretical kinds postulated by scientific theories) and, most importantly, are sources of scientifically compelling answers to 'what is' questions about natural kinds. What better answer to the question 'what is water?' than 'H_2O'. In truth, I suspect that most scientists advocating "scientific" definitions of life are thinking of them along these lines. As discussed below, however, nonstandard definitions of natural kinds cannot be construed as theoretical identity statements.

To begin with, a theoretical identity statement, such as 'water is H_2O', represents an empirical discovery made in the context of a fairly mature, widely accepted scientific theory for understanding and exploring a domain of natural phenomena. Unlike popular definitions of life, theoretical identity statements do not *precede* the development of such theories. As the many competing definitions of life underscore, there currently exists no widely acknowledged, unifying theoretical framework for life. Scientific definitions of life tend to reflect the diverse backgrounds of their authors, with biochemists (e.g., Oparin 1964) emphasizing biochemical aspects of life in their definitions, physicists (e.g., Schrödinger 1944) emphasizing thermodynamic aspects of life in their definitions, and computer scientists (e.g., Langton 1989) emphasizing informational aspects of life in their definitions. The fact that a definition of life draws upon theoretical ideas from a widely accepted, more general, scientific theory does not do much to enhance its plausibility, for the scientific success of the theory concerned rests upon applications to natural phenomena *other than life*. The claim that such theories not only encompass but, most importantly, can be used to distinguish living from nonliving things is highly speculative, and as the heated, discipline driven, debates over definitions of life underscores, very controversial.

It is telling that none of the currently popular definitions of life has the scientific authority of a theoretical identity statement such as 'water is H_2O'. Theoretical identity statements represent *empirical discoveries* in the context of well-established scientific theories. They are not fundamental laws of scientific theories, nor are they deducible from the fundamental laws of scientific theories. However unlikely it may seem, scientists could discover that phenomenal water is not H_2O, just as nineteenth century experimental chemists discovered that Dalton was mistaken in thinking that water is HO. Moreover, labeling a preferred definition of life as "tentative" or "empirically revisable" does not alleviate this problem. In the absence of a widely accepted theoretical framework for exploring alternative possibilities for life, there is little rationale for revising or rejecting a definition of life in light of additional empirical evidence. This helps to explain how someone defending a nonstandard definition of life can get away with simply dismissing counterexamples such as mules or individual organisms as "secondary" forms of life (e.g., Bedau 1998, 129); in the context of a stand-alone nonstandard definition of life, what *grounds* (other than appeals to human intuition and concepts) are there for taking them seriously? The situation is exacerbated when one reflects that, as discussed in Chapter 5, there are compelling theoretical and empirical reasons for believing that familiar life may be unrepresentative of life considered generally. Anyone using one of the currently available definitions to guide the synthesis of life in the laboratory or a search for extraterrestrial life is likely to produce or find only what he or she is looking for. In short, a definition of life is far more likely to blind us to unfamiliar forms of life than to facilitate their discovery.

This brings us to an even more serious difficulty with interpreting a nonstandard definition of life as a theoretical identity statement for life. Despite the appellation, few theoretical identity statements express genuine identities and, as a consequence, they do not provide necessary and sufficient conditions for membership in a natural kind. The claim that water is H_2O provides an especially salient illustration. At first glance, it seems to express an identity (of phenomenal water with the molecule H_2O). A little reflection, however, reveals that this cannot be the case. A single molecule of H_2O lacks temperature and pressure. In the context of modern chemical theory, this is a problem because the unique triple point of water, a distinctive combination of pressure and temperature at which its gas, liquid, and solid phases coexist in equilibrium, distinguishes it from all other chemical substances; different (pure) chemical substances have unique triple points. Furthermore, many of the most distinctive chemical properties of water (e.g., remaining a liquid over a wide range of temperatures, expanding when frozen, being a good solvent) result from secondary structures (dimers, trimers, etc.) produced by weak hydrogen bonds among H_2O molecules. Finally, the purest samples of phenomenal water are not composed of *just* H_2O molecules. H_2O molecules invariably dissociate into the ions H^+, OH^-, and H_3O^+ when combined. In addition, many other chemical substances, such as hydrochloric acid, include large quantities of H_2O; under normal conditions of pressure and temperature, hydrochloric acid vaporizes at room temperature and pressure unless it is mixed with more than 60% water. In short, while phenomenal water is predominately made up of H_2O molecules it cannot be literally identified with them. Moreover, no exact number or percentage of H_2O molecules is required for a phenomenal substance to qualify as water. The best that can be said is that *being H_2O* is the most salient theoretical property of water *within the broader context of modern chemical theory.*

I have not of course proven that it is impossible to supply necessary and sufficient conditions for being water within the framework of modern chemical theory. In this context, it is somewhat amusing to consider the lengths to which Putnam went in trying to formulate a theoretical identity statement for water in the face of a seemingly endless supply of counterexamples.[15] He found that it was far more difficult than he initially supposed. His penultimate version (Putnam 1983) is as follows:

[Water is] ... a quantum mechanical super-position of H_2O, H_4O_2, H_6O_3, ... plus D_2O, D_4O_2, ...

(p. 63)

And he (somewhat tellingly) muses that:

These examples suggest that the "essence" that physics discovers is better thought of as a sort of *paradigm* that other applications of the concept ("water", or "temperature") must *resemble* than as a necessary and sufficient condition good in all possible worlds.

(p. 64)

In short, there is little support for the claim that scientific theories supply necessary and sufficient conditions for the natural kinds that they subsume. It is also worth noting that chemists do not share Putnam's urgency about supplying necessary and sufficient conditions for water or, for that matter, any other chemical substance. They do not share his urgency because they do not need to fix the extensions of terms for chemical kinds in order to reason successfully about them. Chemists differentiate among chemical substances by exploiting different parts of chemical theory (an amalgam of chemical thermodynamics, molecular theory, and quantum mechanics) and using methods and tools that are constantly changing with advances in technology. In short, scientific theories are much more fluid and open ended in their classifications than Euclidean geometry. I will have more to say about this in the following section, as well as Chapter 4.

3.4.2 Nonstandard Definitions Do Not "Define" Scientific Theories

The question remains: Could a nonstandard definition of life be interpreted as encapsulating a preliminary theory of life? This is the approach taken by some

---

[15] As Section 2.3 discussed, Putnam rejected the claim that theoretical identity statements are *bona fide* definitions (in the logician's sense). Initially, however, he treated them as empirically confirmed identity statements for the natural kinds subsumed by scientific theories. Accordingly, the early Putnam would have been somewhat sympathetic to the concept of a nonstandard definition, although as a logician, he undoubtedly would have resisted pressure to dub them 'definitions'.

philosophically sophisticated advocates of definitions of life (e.g., Bedau) who interpret them as theoretical definitions in the sense of the semantic conception of scientific theory. Just how plausible is this strategy?

There are several problems. The theoretical definitions of the semantic conception consist of logical conjunctions of fundamental laws. They define scientific theories as integrated wholes; they do not supply separate sets of necessary and sufficient conditions for the natural kinds that a theory subsumes. Consider classical mechanics, one of the few scientific theories that it is plausible to think could be encapsulated by a theoretical definition. A conjunction of Newton's three "laws" of motion does not state necessary and sufficient conditions for being a body in motion even though this is the class of phenomena targeted by the theory. Instead, classical mechanics provides a theoretical framework, consisting of theoretical concepts (inertial mass, position in space and time), as well as concepts more closely related to observation (velocity and acceleration), coupled with three core theoretical principles for relating and integrating them, for purposes of prediction and explanation. At best a conjunction of Newton's three laws can be said to define '[theory of] classical mechanics,' which underscores just how different theoretical definitions are from nonstandard definitions.

The holistic character of theoretical definitions does not of course prove that one could not deduce necessary and sufficient conditions for being a 'body in motion' from a conjunction of Newton's three laws of motion. But as we have seen, this is not true in general for scientific theories. What was revolutionary about Newton's theory was not that it fully settled the extension of the expression 'body in motion' but that it greatly enhanced the ability of his contemporaries to predict and explain the motion of bodies. There is no guarantee that supplying necessary and sufficient conditions for membership in the class of items consisting of bodies in motion will yield an analysis having this sort of empirical significance, and as the case of modern chemistry discussed earlier underscores, good reasons for believing that this kind of empirical significance can be achieved without supplying necessary and sufficient conditions for being a body in motion. The point is that an intuitively appealing classification scheme does not guarantee predictive and explanatory success of the sort sought by scientists.

In this light, consider Bedau's (1998) definition-theory of life: In his words,

[D_B] So, the general form of my theory of life can be captured in this definition:
    X is living *iff* [if and only if]

1. X is a supply adapting system, or
2. X is explained in the right way by a supply adapting system.

(p. 128)

where, according to Bedau, a "supply adapting system" engages in "the ongoing production of significant adaptive novelty" (p. 127). Bedau denies that D_B is a definition in the classical sense. When pressed about what sort of definition it is, he claims that D_B is a theoretical definition in the sense of the semantic conception of scientific theory (personal communication). It is obvious, however, that D_B lacks the form of a theoretical definition. As the use of the biconditional (iff) indicates, D_B supplies necessary and sufficient conditions for being a member of the (putatively) natural kind life. In contrast, chemical theory does not supply necessary and sufficient conditions for any of the enormous number of natural kinds (chemical substances) that it encompasses.

Bedau does not, however, claim that D_B constitutes a mature theory of life. He emphasizes its tentative and revisable nature, which suggests another possibility for interpreting D_B. As mentioned earlier, some proponents of the semantic conception treat linguistic formulations of theories as if they were merely heuristic devices – as if one could directly define a scientific theory nonlinguistically, along the lines of a mathematical theory, as a model theoretic structure. Can D_B be interpreted as providing a preliminary sketch of a model for life?

Alas, no. D_B does not describe a model of any sort. It explicitly delimits a *category*: the class of living things. Bedau is not the only philosopher who conflates theories and (nonstandard) definitions in this way. Ruiz-Mirazo et al.'s (2004) "definition" (their term) of life provides another illustration:

[D_R]   "Life" – in the broad sense of the term – is a complex collective network made out of self-reproducing autonomous agents whose basic organization is instructed by material records generated through the evolutionary-historical process of that collective network.
(p. 339)

It is clear from their discussion that they intend D_R as a preliminary theory of life. Like D_B, however, the term 'life' is explicitly treated as if it were the definiendum of a definition. D_R provides necessary and sufficient conditions for falling into the class of living things. This is not what one would expect from a preliminary scientific theory of life, whether interpreted as a deductively closed axiom system (syntactic conception) or as a set theoretic structure (semantic conception); for as we have seen, on neither conception does a scientific theory supply necessary and sufficient conditions for the natural kinds that it subsumes. Labeling a nonstandard definition of life as "tentative and revisable" does not alleviate this difficulty. It serves only to blur important ways in which scientific theories differ from definitions.

The primary function of all definitions (traditional and nonstandard) is classification. Scientific theories, in contrast, have a variety of functions other than classification. As mentioned earlier, these functions include explaining and predicting phenomena falling under a theory, but they also include guiding and constraining the search for relevant evidence and interpreting data once it is acquired. The ability to come up with an intuitively appealing classification scheme – to sort things into distinct, highly plausible, categories – does not guarantee the capacity to realize these other functions, which are far more critical to the scientific success of a theory. Indeed, as discussed in Chapter 4, only a small subset of classification schemes is capable of supporting the latter. The credibility of a classification scheme depends upon our current scientific beliefs about the phenomena concerned. If those beliefs are wrong then the classification scheme is unlikely to be scientifically fruitful. To cut a long story short, the scientific project of defining life rests not only upon misconceptions about a close logical relationship between definitions and scientific theories but also upon ignoring the important functions served by scientific theories.

3.5 Concluding Thoughts

Reconstructing scientific theories as either deductively closed axiom systems (syntactic conception) or set theoretic structures (semantic conception) provides a much needed standpoint for comparing and contrasting their structure with that of maximally informative definitions, which have a clear-cut, very simple structure. As Section 3.4 discusses, from the perspective of defenders of nonstandard definitions of life, the results are disappointing. Admittedly, under the semantic conception, which still dominates philosophical thought about the structure of scientific theories, an entire scientific theory can be compressed into a theoretical definition. The latter does not, however, provide necessary and sufficient conditions for the natural kinds encompassed by the theory; a theoretical definition "defines" the components of a theory as a tightly integrated, indivisible unit. Furthermore, as Putnam's thwarted efforts to provide necessary and sufficient conditions for water in the context of quantum chemistry underscores, one cannot logically infer nonstandard definitions for natural kinds from the theories that subsume them. Scientific theories are far more open ended and fluid than definitions, which is what allows them to be applied to novel and unanticipated empirical phenomena. In sum, it is a mistake to think that that one could logically infer a nonstandard definition of life from a (tentative or mature) universal theory of life or that a (tentative or mature) scientific theory of life could be compressed into a nonstandard definition *of life*.

It is also a mistake to hope that the scientific project of defining life can be revived under the newer pragmatic conception of scientific theories.[16] Although there are different versions of the pragmatic conception, they all reject the claim that scientific theories have a one-size-fits-all, uniform, formal

---

[16] For a good review of the pragmatic conception of scientific theories, see Winther (2015).

(logicomathematical) structure. Indeed, many versions of the pragmatic conception include informal aspects of scientific reasoning thought to play critical roles in scientific modeling and theorizing, as part of the structure of scientific theories. As a result, the pragmatic conception is even less able than the syntactic and semantic conceptions to support the claim that there is a close structural relationship between scientific theories and definitions.

Finally, some philosophers (e.g., Bich and Green 2017; Knuuttila and Loettgers 2017) reject the received view (among philosophers and logicians) that a central function of definition is classification. They contend that it is a mistake to require that a "definition" of life supply necessary and sufficient conditions, thus hoping to undercut the arguments of Cleland (2012), Cleland and Chyba (2002, 2007), and Machery (2012) against the scientific project of defining life. On their proposals, which are very similar, certain "definitions" of life used by scientists – which they dubb "operational" (Bich and Green), and "theoretical," "transdisciplinary," and "diagnostic" (Knuuttila and Loettgers) – do not aim at distinguishing life from nonlife but nevertheless play important theoretical and methodological roles in scientific practice.

There are several problems with these proposals. First, the fact that some scientists refer to fuzzy, incomplete assertions of the sort concerned as "definitions" does not mean that they are definitions. Moreover, as we have seen, 'operational definition' (discussed in Section 2.3.1) and 'theoretical definition' (under the semantic conception) have long established, technical meanings in philosophy of science, and the ways that these terms are being used by Bich, Green, Knuuttila, and Loettgers does not conform to those meanings. So why even call them "definitions"? Why change the meaning of the word 'definition' instead of coining a new term for them? The reason, I suspect, is somewhat ironic: Definitions are very appealing because they promise clear-cut answers to questions such as 'what is life?' and they do this *in virtue of supplying necessary and sufficient conditions* (for life). In this spirit, one might generously interpret the things that they are calling definitions as half-way houses in the quest for a full-fledged definition of life, but doing so would reinstate the notion that authentic definitions supply necessary and sufficient conditions for the terms that they define.

There is another, even more serious, problem with the claim that the aim of many scientific "definitions" of life is not classificatory. It just is not true that most scientists who speak of "defining" life are not interested in distinguishing between life and nonlife. Such talk occurs most frequently among astrobiologists, ALife researchers, and scientists investigating the origins of natural life. This is not an accident. ALife and origins of life researchers are concerned with the question of when a nonliving system (whether a simulation, robot, or ensemble of molecules) makes the transition to a primitive living thing, and this is a question about classification. Similarly, astrobiologists want to be able to differentiate novel forms of life from nonliving extraterrestrial phenomena. In contrast, biologists concerned with exploring similarities and differences among known Earth organisms for example, whether eukaryotes should be classified under domain Archaea or how birds differ from reptiles are much less interested in defining life; they are not faced with the challenging classificatory problem of distinguishing novel forms of life from unusual nonliving systems.

But now we seem to be faced with a conundrum: How can scientists search for unfamiliar forms of life, whether here on Earth, a shadow biosphere (Chapter 9), or elsewhere in the universe (Chapter 8), in the absence of a definition or universal theory of life? Chapter 8 proposes a nondefinitional solution: Use tentative (versus defining) criteria to search for potentially biological anomalies, as opposed to trying to prematurely decide the status of a puzzling phenomenon *by definition*. Once a suspiciously biological anomaly is identified it becomes a candidate for further, more in-depth, scientific investigation for life. While this strategy will not decisively settle the question of whether a puzzling phenomenon represents an instance of unfamiliar life, it has an important advantage over a definitional approach: It will decrease the likelihood of misidentifying a truly novel form of life as just another mysterious abiological phenomenon, and it will do just as well at identifying forms of life that closely resemble our own as *bona fide* living things.

4

How Scientific Theories Develop

4.1 Overview

What factors impede the development of successful scientific theories? How can the development of such theories be facilitated? These questions arise independently of which conception of scientific theory one endorses. They are especially important for biology since we currently lack a scientifically fruitful, universal theory of life; as many biologists are fond of admonishing, "give me a general principle of biology and I'll find an exception." At best (*assuming* that life has a universal nature, which is not certain) we are still in the earliest stages of formulating such a theory. For as the next chapter (Chapter 5) explains, recent advances in biochemistry and molecular biology have established that familiar Earth life provides just a single example of life. Biologists have also discovered that complex multicellular eukaryotes are highly specialized, biologically fragile, latecomers to our planet. Yet (as discussed in Chapter 1) the latter, especially animals and plants, have served as prototypes for biology since the time of Aristotle. In a nutshell, in addition to generalizing on the basis of a single example of life, we have been seeking universal principles for life from an unrepresentative subsample of it.

Viewed from this perspective, it is hardly surprising that biologists have yet to formulate a scientifically fruitful, truly general, theory of life. As an analogy, one would not expect an extraterrestrial biologist to be able to formulate a universal theory of mammals on the basis of observations of a single example, say, zebras. For purposes of generalization, our extraterrestrial biologist is more likely to focus on their ubiquitous stripes than on their mammary glands; after all, only half of zebras (females) have the latter. Yet as any human biologist will tell you, having mammary glands is the key to understanding how mammals differ from other animals, such as birds and fish, and understanding this requires experience with additional examples of mammal. The point is a single example of a natural phenomenon cannot reveal which of its characteristics are "accidental" and which are fundamental. Nonetheless, this distinction is crucial to the success of a scientific theory. One cannot construct a successful *general* theory for a domain of phenomena on the basis of characteristics contingent upon the circumstances in which a *particular* example arose. Indeed, as Section 4.3 discusses, even when one has several independent examples of a phenomenon it is often difficult to discriminate the "essential" from the fortuitous.

These considerations suggest that it will be very difficult, if not impossible, to begin the process of developing a universal theory of life in the absence of examples descended from an alternative origin of life. For how else can one distinguish basic features of life from those contingent upon chemical and physical conditions on the young Earth during the origin and early development of known life? Fortunately, the situation is not as hopeless as it sounds. Theories do not spring full-blown from the minds of scientists. They develop in fits and starts. It is not too early to begin the process of formulating and exploring tentative theoretical frameworks for a universal biology *so long as one keeps in mind that they are a work in progress*, subject to revision, augmentation, and even elimination in light of new empirical discoveries and theoretical developments.

As Section 4.2 explains, theory construction begins with the development of an *ontology* (set of core theoretical concepts) for formulating generalizations about a seemingly unified domain of natural phenomena. Whether one can formulate fruitful generalizations for purposes of prediction and explanation, among other scientific activities, depends upon selecting an auspicious ontology; select the wrong basic concepts and they will not be forthcoming. Newton's second law of motion ($F = ma$), for instance, presupposes the theoretical concepts of mass and position in space and time as basic; force, velocity, and acceleration are defined in terms of the former, and hence are not basic. Newton's predecessors were unable to formulate universal laws of motion because the basic theoretical concepts that they were working with (impetus, occupied volume, and weight) could not support them. Unfortunately, as Section 4.4 discusses, in the context of examples from the history of science, scientists have a tendency to settle quickly on an ontology and are reluctant to abandon it even when faced with persistent failure to find empirically reliable, general principles. It is thus important to resist becoming prematurely wedded to an ontology – to be open to the possibility that the problem is not (as scientists typically assume) a failure to come up with the correct fundamental principles or (as many pluralists contend) the intrinsic disunity of the phenomena under investigation, but instead a defective theoretical framework of basic concepts for formulating general principles. The need to resist becoming prematurely committed to a set of basic theoretical concepts is especially important for the fledgling program of universal biology.

4.2 How Scientifically Fruitful Ontologies Develop: Content Matters

At the heart of a mature scientific theory is a set of fundamental theoretical concepts and principles for reasoning about the phenomena that it subsumes. The theoretical concepts, which comprise the ontology of the theory, designate what are postulated to be basic entities. Basic entities may include objects, properties (including nonlogical relations), processes or mechanisms. The theoretical principles of the theory identify empirically significant patterns or regularities holding among the entities designated by the core theoretical concepts. In physics, the theoretical principles ("fundamental laws") take the form of deterministic or probabilistic quantitative generalizations. It should not be assumed, however, that the foundational principles of every scientific theory must be like those of physics. Qualitative principles – which might, for instance, describe causal relationships among basic mechanisms – may be more fruitful for understanding some aspects of nature. Theoretical principles represent conjectures about how the theoretical categories postulated by the ontology are related to each other and ultimately – through, for example, correspondence rules, under the syntactic conception, and theoretical hypotheses, under the semantic conception – to the familiar world of phenomenal experience.

Unhappily, nature does not come pre-structured into a set of propitious theoretical concepts for formulating scientifically fruitful generalizations. In order to evaluate the adequacy of a scientific theory, the target phenomena must be conceptually carved into basic categories, which presupposes privileging certain aspects over others as fundamental. The semantic conception of scientific theory (Section 3.3) provides a salient illustration. The set theoretic structure (family of models) constituting a theory provides a theoretical framework for organizing a chunk of nature into "replicas" for purposes of description, explanation and prediction; otherwise one cannot compare the theory to the world, that is, establish relations of isomorphism, homomorphism, or (informal relations of) similarity between the family of models individuating the theory and the phenomena of interest. In essence, the ontology of a scientific theory functions like a cookie cutter for extracting a replica from a target phenomenal system.

When scientists lack a mature theory for a domain of natural phenomena, as they do in the case of life, the direction of analysis reverses and serious questions arise. How does one carve up (in William James's (1890, 488) colorful words) the "blooming, buzzing confusion" of a raw portion of phenomenal nature for purposes of scientific reasoning? The number of conceptual possibilities is enormous. In the absence of a widely accepted theoretical framework there is little basis for preference except for the pre-theoretical perceptual kinds already available through natural language. Viewed in this light, it is hardly surprising that the four basic elements of the ancient Greek philosophers (water, air, earth, and fire) correspond to familiar perceptual kinds.

On the other hand, pre-theoretical categories, such as water and fire, are what a modern scientific theory is supposed to explain via a more basic theoretical classification scheme supporting scientifically fruitful, unifying principles. Whether one is able to discover empirically powerful, foundational theoretical principles for reasoning about a domain of natural phenomena critically depends upon the ontology selected. Most ontologies produce dead ends, regardless of how well (from an intuitive perspective) they seem to partition the phenomena of interest into categories, which underscores a point made in the last chapter: Unlike the case with definitions, carving a domain of natural phenomena into categories is only one function of a scientific theory. As the history of science reveals, what really matters is how *fruitful* the categories are for purposes of scientific practice, and the latter depends on whether they can support empirically powerful, comprehensive principles for explaining, predicting, manipulating, interpreting, and otherwise exploring the phenomena of interest.

The history leading up to Newton's theory of motion provides a revealing illustration of how an unpropitious ontology can frustrate the development of a scientifically fruitful universal theory.[1] Up until around the time of the Copernican revolution, Aristotle's ideas dominated thought about motion. Aristotle sharply distinguished celestial motion from terrestrial motion on the basis of observation. Celestial bodies (stars, planets, and moons) are constantly in motion. Like living things, they seem to move under their own power, without the action of an external cause. Inanimate terrestrial bodies, such as stones, on the other hand, seem to move only when caused to do so. Aristotle concluded – *On the Heavens* Bk 1.2–1.3 and *Metaphysics* IX.8 (e.g., trans. Barnes 1984) – that there is a fundamental difference between celestial motion and terrestrial motion. He theorized that this difference has to do with the composition of the bodies involved. Celestial bodies are not made up of the same stuff as terrestrial bodies. There are passages (e.g., *On the Heavens* Bk XII, especially 292^{a17}–293^{a1}) in which Aristotle even speaks of celestial bodies as living entities on the grounds that they are self-moving; it will be recalled that, in Aristotle's biology, the latter is a distinctive functional capacity of animal life (Section 1.2). The important point for our purposes, however, is that, for Aristotle, celestial bodies are not composed of the four basic terrestrial elements, namely, water, air, earth, and fire. They are composed of an ethereal fifth element.

Aristotle theorized that the natural state of inanimate terrestrial bodies is rest since, unlike celestial objects, they invariably cease moving (*Physics*

---

[1] It is of course beyond the scope of this chapter to provide a detailed account of the historical development of Newton's theory; see, e.g., Falcon (2005) and Cohen and Smith (2002) for more historical detail.

Bk VIII.4–VIII.10). What therefore needs to be explained is how an object, such as a stone, begins to move and stays in motion. The motion of inanimate terrestrial bodies requires an external cause. Projectile motion posed a serious challenge for Aristotle's account because nothing observed keeps a projectile in motion after it is launched. It appears to move under its own power before coming to rest. In the late Middle Ages the concept of "impetus" (an internal, motive cause for keeping an object in motion) was advanced by neo-Aristotelians as a solution to this pressing problem.

With the advent of the Copernican revolution, the concept of impetus fell into disfavor. Among other things, it could not explain an apparent contradiction between the observed behavior of falling bodies and the Copernican claim that Earth revolves on its axis. On the assumption that Earth rotates from west to east, a ball dropped from a tower should fall to the west of the tower rather than straight down. To resolve these and other difficulties, Johannes Kepler broke with the neo-Aristotelians and entertained the (then) radical idea that matter is inherently "inert," i.e., unable to move by itself (Cohen 2002); he coined the theoretical term "inertia" for this property. On Kepler's view, it is not motion *per se* but rather *change* in motion (which includes ceasing to move, as well as beginning to move) that requires causal explanation. In other words, instead of following Aristotle and asking what keeps a material body in motion, Kepler turned the question around and asked what causes a moving body to stop.[2] Reframing the question in this way set the stage for Newton's highly fruitful, truly universal, theory of motion.

Newton refined Kepler's suggestion into the principle of inertia (enshrined in his first law of motion). The principle of inertia holds that a material body is indifferent to its state of rest or uniform motion. In this manner, Newton was able to subsume an astonishing variety of *prima facie* unrelated perceptible phenomena – for example, the motion of projectiles, pendulums, colliding billiard balls, falling bodies, and, most importantly, celestial bodies – under four basic laws, three of motion and one of universal gravitation, by appealing to the propitious theoretical concepts of mass and inertia. Newton identified mass and inertia (Cohen 2002), which gave rise to the concept of inertial mass. The concepts of mass and inertia are notable for their distance from the phenomena of everyday human experience. Yet without them Newton could not have completed the unification of celestial motion and terrestrial motion begun by Copernicus in the sixteenth century.

It is instructive to compare Newton's concept of material body with those of his most famous predecessors, the giants on whose shoulders he claimed to stand. Descartes identified quantity of matter with purely geometrical characteristics,

---

[2] For more detail on the historical development of the concept of inertia from Galileo to Newton, see Chalmers (1993) and Westfall (1977).

namely, occupied volume. On Newton's theory of motion, however, bodies of the same size and shape (e.g., 1 inch spheres of iron and glass) may have different masses, depending upon their inertia (resistance to applied forces); this difference is measurable through the second law of motion ($F = ma$) by solving for mass, m, where F stands for an applied force and a stands for the resultant acceleration. Galileo, on the other hand, identified quantity of matter with weight, a view closer to Newton than to Descartes, who took weight to be an accidental property of matter. Newton nonetheless broke with Galileo in carefully distinguishing weight (w) from mass (m) in his law of terrestrial free fall (viz., $w = mg$, where $g = 32$ feet/sec^2 or $g = 9.81$ meters/sec^2). On Newton's law of universal gravitation, $F = G(m_1 m_2 / r^2)$, where G is the universal gravitational constant and r is the distance between the two masses, m_1 and m_2, objects with the same mass (m_1) have significantly different weights on, for instance, Earth and Mars – which have different constants of acceleration because they have different masses (m_2) – and objects with quite different masses have the same (namely, zero) weight in empty space. The only differences between inertial mass, in Newton's three laws of motion, and gravitational mass, in his law of universal gravitation, are the methods used to measure them. Because these methods yield the same quantitative result, Newton and his successors concluded that inertial mass and gravitational mass are the same. In essence, for Newton, weight is a perceptible manifestation of inertial mass. Newton was thus able not only to explain phenomenal properties associated with material bodies, such as bulk (occupied volume) and weight, but also unify terrestrial and celestial motion by appealing to the intangible concept of inertial mass.

But what exactly does the term 'mass' designate? The answer is not at all clear. Unlike occupied volume and weight, mass is not directly perceptible. Mass is sometimes characterized as a measure of inertia. But this reply does not help much since inertia (*qua capacity* to resist a change in motion if acted on by a net force) is not directly observable either. Similarly, forces are not observable except through changes in the motion of bodies. Nonetheless the highly abstract concepts of mass, inertia, and force are indispensable to the success of Newton's theory of motion. Without them he could not have integrated and unified all motion in the universe under just four fundamental laws.

The failures of Aristotle, Galileo, Descartes, and many others to formulate empirically reliable, universal principles of motion can at least in part be blamed on inadequate ontologies. Earlier concepts of matter as bulk, impenetrability, and weight were incapable of supporting such principles even though they are more accessible to human observation than the rarefied concept of inertial mass. Newton's ontology, in contrast, was able to subsume all motion under four laws of motion. In addition, his new theory was able to explain the characteristics of material bodies previously taken as fundamental, sometimes, as discussed below, with startling results.

On Newton's theory, occupied volume is not necessary for being a material object. It is in principle possible for a material body to be unextended (a point mass). Whether entities as strange as point masses actually exist, as opposed to representing unrealizable theoretical idealizations, is an interesting question. But because the theoretical framework of modern physics follows Newton in permitting them, they have been conjectured to exist. Black holes provide a salient illustration. A black hole develops when an extremely massive star undergoes complete gravitational collapse. It is widely believed that it is physically possible for a massive star to collapse symmetrically to a point singularity where all the mass of the original star is concentrated in a mathematical point; having zero volume, a point singularity is infinitely dense. Astronomers have discovered a number of black holes, although there is disagreement as to whether the mass they contain is concentrated in a point singularity, as opposed to an extremely small volume. To cut a long story short, the empirical possibilities that scientists are willing to entertain – the research projects that they are willing to undertake – depends upon their choice of theoretical framework. That is to say, the ontology that one selects for systematizing a domain of natural phenomena not only determines whether one can formulate explanatorily and predictively powerful generalizations about it but also which research projects are considered worthy of scientific investigation.

To wrap up, the selection of a promising set of basic theoretical concepts is critical to the development of a successful scientific theory. Select an unpropitious ontology and empirically and theoretically fruitful generalizations for scientific practice will not be forthcoming. As discussed below, however, coming up with a promising ontology is not easy.

4.3 The Goldilocks Level of Abstraction

While abstracting away from detail is clearly necessary for successful generalization, it has become increasingly clear that formal model theoretic analyses of scientific theories (see Section 3.3) carry the process too far; for a review of this literature, see Frigg and Hartmann (2012). Theories having different ontologies (commitments to different types of *nonlogical* objects, properties, processes, or mechanisms) may nonetheless share the same, stripped down, logicomathematical structure. But it does not follow that these theories have the same predictive, and explanatory power, among other important scientific virtues, vis-à-vis a given domain of natural phenomena. To appreciate this point, consider a formal structural description of a chemical substance such as water and ask whether one could infer the distinctive chemical and physical properties of water from it. One cannot. This is not a new point. It is closely related to the question of whether there is a difference between a computer simulation and the proverbial real McCoy. Purely formal systems have purely formal consequences. But despite the musings of some theoretical physicists and philosophers,[3] the world of nature does not seem to be a purely formal (informational) entity. Flesh and blood organisms can no more digest the pretend sugars produced by a digital simulation of photosynthesis (however detailed the structural description) than they can quench their thirst with simulated water; we will return to these issues in Chapter 7. As Mattingly (2005) counsels, ". . . not all the content of a given theory can be shoe-horned into the structure provided by the semantic conception" (p. 367), and content matters for purposes of scientific explanation and prediction.[4]

Considerations such as these have led a growing number of philosophers sympathetic to the semantic conception – for example, Cartwright (1983), Downes (1992), Griesemer (1990), Lloyd (1988), Morrison (1999), Weisberg (2013), and Wimsatt (1987) – to emphasize the *content* of a theory (what it says about the world) while retaining commitment to a model-based account of scientific practice. Some but not all of these philosophers identify theories with models. All of them, however, view models as indispensable tools for doing science. Instead of seeking a one-size-fits-all formal analysis of scientific models, these philosophers focus on the diverse ways in which the models actually used in science function in scientific practice. For our purposes here, however, the crucial point is that the scientific fruitfulness of a theory depends upon its content as well as its structure. Successful theories presuppose ontologies of a very special sort, namely, those capable of divvying up a target phenomenal system into categories supporting empirically and theoretically powerful unifying principles for scientific practice.

To unify a domain of natural phenomena, a scientific theory needs to be universal but it need not be exceptionless. The two are often assumed to go together but, strictly speaking, no scientific theory is truly exceptionless, and this includes those of physics (Cartwright 1983, 1989; Earman et al. 2002). The supposition that the laws of physics are exceptionless is maintained through *ad hoc* appeals to *ceteris paribus* (all other things being equal or constant) conditions for explaining what otherwise seem to be violations of law; ostensible

---

[3] The most well-known physicist is Max Tegmark (1998), who contends that the universe is ultimately (when properly understood) a purely mathematical object; he calls this the "mathematical universe hypothesis." Philosophers Don Ross and James Ladyman (2010) also seem to be defending a structuralist view (appropriately dubbed "information-theoretic structural realism") along these lines.

[4] I am not claiming that formal models of natural processes are scientifically useless. As Elliot Sober (1991) observes, some aspects of nature do seem to be purely formal (he cites biological fitness as an illustration), and as a consequence, computational models provide excellent tools for exploring them. But as Sober also argues, not all characteristics of natural phenomena are purely formal: In his words, "the digestive process, *per se*, is not computational" (p. 762).

exceptions are handled by appealing to scientifically plausible differences (which may not be fleshed out or even well understood) among phenomena allegedly falling under them. In other words, the exceptionlessness of the laws of physics is primarily an article of faith. Admittedly, as Sandra Mitchell (2000, 2002) points out, the laws of physics are especially successful at combining universality in scope with near exceptionlessness. As she puts it, they have more "stability under contingency" (varying conditions) than the basic principles of other scientific disciplines. But as her reference to contingency underscores, their exceptionlessness is not perfect.[5]

For a principle to unify a domain of natural phenomena, only universality in scope is required. In scientific (versus mathematical) theories, there is typically a tradeoff between the unifying power of a generalization and its ostensible exceptionlessness. Unification depends upon highly selective idealizations – on abstracting away from the messy contingent details of *particular* phenomena to lay bare patterns capable of being generalized to all phenomena *of that kind*. The "ideal gas law" ($PV = kT$, where P stands for pressure, V for volume, T for temperature, and k is a constant for a given amount of gas) provides a salient illustration. It applies to the state of a hypothetical ideal gas, and is a good approximation to the behavior of many real gases under a wide variety of conditions. Because it neglects the sizes of gas molecules and intermolecular attractions and repulsions, however, no real gas actually satisfies it, although some come close to doing so. As a consequence, the ideal gas law is riddled with (real world) exceptions even though it is universal in scope (applies to all gases). The reason that these exceptions are not viewed as challenging the universality of the law is that they fall outside of it. They are explicable in terms of characteristics, such as the size, shape, and charge of gas molecules, that vary among different gases. The ideal gas law abstracts away from such characteristics for purposes of generalizing across all gases.

Ostensible exceptions to scientific principles are not always explicable in terms of contingencies but it is not easy to tell when this is the case. Newton's law of universal gravitation provides a good illustration. It says nothing about the number and locations of planets and moons in solar systems, raising the question of whether an ostensible exception to the law is the result of contingent facts about *our* particular solar system – the gravitational influence of an as yet undiscovered body (whose existence or nonexistence is independent of the law) – or represents a defect in the law itself. From the time of its inception, Newton's law of gravitation

---

[5] One cannot of course exclude the possibility that these exceptions merely reflect the inadequacy of our current knowledge about the universe – that an extraordinarily complex, universal, exceptionless theory of everything is waiting in the wings (so to speak) if we are just clever enough to formulate it. I will have more to say about this possibility in Section 4.5, where I promote "the monist stance." For our purposes here, however, my point is epistemological: None of our current scientific theories is truly exceptionless.

faced many exceptions, most of which were resolved with the discovery of additional planetary bodies. Uranus's deviant orbit, for instance, was fairly quickly explained in terms of the gravitational influence of a previously unknown planet, Neptune. Mercury's orbit, on the other hand, remained a mystery until the advent of Einstein's general theory of relativity. The deviation in Mercury's orbit was not due to contingent facts about the distribution of planetary bodies in our solar system. It represented a genuine counterexample to Newton's law of universal gravitation. Pluralists who reject the program of universal biology on the grounds that every principle of biology thus far entertained has exceptions, conflate universality with exceptionlessness. They fail to appreciate that even the most reliable scientific principles face exceptions resulting from poorly understood contingencies, as opposed to theory defeating counterexamples.

For purposes of this chapter the take home point is that one cannot unify a domain of phenomena without abstracting away from detail, and this means that the ontologies of scientific theories inevitably open the door to exceptions in the theoretical principles in which they figure. It would thus be a mistake to condemn a fledgling theory of life merely on the grounds that it is not exceptionless. This contrasts with the definitional approach, which as underscored by the obsession with eliminating counterexamples (see Chapter 2), mistakenly aims for *both* universality and exceptionlessness. The important question is not whether a scientific theory's principles are truly exceptionless but instead whether the exceptions can be explained in terms of plausible contingencies holding in particular cases.

The development of modern chemistry provides an illustration of both the need for and difficulties involved in hitting the right level of abstraction. Despite its universality, Newton's theory of motion could not explain the behavior of different chemical substances. It was pitched at too abstract a level of analysis. The effects of mechanical forces are negligible compared to the chemical reactivity and miscibility of different chemical substances when combined. Until the late eighteenth century, chemists were still under the influence of Aristotle's two-thousand-year-old theory of terrestrial matter, which held that all chemical substances consist of four basic elements: air, water, earth, and fire. With the exception of fire (whose status as an element was no longer secure), these elements correspond roughly to what we now call "phases" of matter: gas, liquid, and solid. The tendency to associate water with a liquid in ordinary, everyday discourse may be a vestige of this ancient view. Aristotle and many of his successors interpreted a change in phase as a change (transmutation) from one kind of substance to another (Needham 2002). When ice melts it transforms into a completely different substance, (liquid) water.

The key to the development of modern chemistry was freeing the concept of chemical substance from that of phase, and this required rejecting Aristotle's ontology for chemical substance (Needham 2002). The first steps were taken in the late eighteenth century when Lavoisier established the law of definite proportions experimentally for water, demonstrating that water consists of two other substances, hydrogen and oxygen (which he could not further decompose), in fixed proportions by weight. A phase-independent concept of chemical substance represented a fundamental change in theoretical framework for chemistry, and set the stage for Josiah Gibbs's application of classical thermodynamics to chemical problems in the late nineteenth century. Gibbs formulated the phase rule $F = C - P + 2$ for a chemical system, where P is the number of phases in thermodynamic equilibrium, C is the number of components (chemical substances), and F is the number of degrees of freedom (e.g., combinations of temperature and pressure). Gibbs's phase rule was crucial to the development of modern chemistry. It provided a new foundation for distinguishing among chemical substances. If a chemical system consists of only one component (a single chemical substance), $C = 1$, and if all three phases (solid, liquid, and gas) of that substance are in equilibrium, $P = 3$, it follows that $F = 0$ (there are no degrees of freedom). In other words, there is only one combination of temperature and pressure at which all three phases of a chemical substance can be in equilibrium. This temperature and pressure is known as a substance's triple point. The concept of the triple point provided a foundation for differentiating "pure" chemical substances from chemical mixtures.

In sum, the most fruitful ontologies support generalizations at just the right level of abstraction for purposes of explanation and prediction. Generalizations that are too abstract may be universal in scope but will be riddled with potential exceptions that are virtually impossible to resolve in a scientifically compelling manner. Generalizations that are too concrete, on the other hand, will be too limited in scope to encompass the full range of related phenomena. One needs to hit just the right – what I dub the "Goldilocks" – level of abstraction for maximizing scope while minimizing (but not eliminating) exceptions.

It is impossible to say in advance what the Goldlilocks level of abstraction is for a phenomenal system of interest. There is no reason to suppose that it is the same for every aspect of nature. As Hilary Putnam (1975, 291–303) observed some time ago, quantum mechanics provides a poorer explanation of why a square peg will pass through a square hole but not a round hole than the plane geometry of rigid bodies. In this case, high level structure matters more than material constitution. In earlier work, I argued that space is best understood as a structural phenomenon (Cleland 1984). Many aspects of nature, however, do not seem to be purely structural. As Chapter 7 argues, there are compelling reasons for thinking that life, which manifests as a complicated chemical phenomenon, is one of the latter.

The Goldilocks level of abstraction is more a zone than a level.[6] There is no guarantee that there is a single unique way of subdividing a target phenomenal system into categories for purposes of successfully generalizing about it scientifically. One cannot eliminate the possibility that there are different ways of carving a given aspect of nature into basic categories, each of which is capable of supporting generalizations that (while different) are equally good at maximizing scope and minimizing exceptions and satisfying other important scientific virtues ("epistemic values"), such as consistency, accuracy, simplicity, explanatory power, predictive power, etc. How many different possibilities there are for carving a given domain of phenomena into scientifically fruitful ontologies is always an open question.[7] In some cases, there may be few. In others, there may be many. But, and this is the important point, there will be many more possibilities for arbitrarily subdividing a phenomenal system into categories that are scientifically ineffectual than there will be for subdividing it into categories that are scientifically fruitful. Not just anything goes.

This brings us to an action point: Avoid premature commitment to an ontology, especially one whose main virtue (like that of contemporary definitions of life) is intuitive appeal (based on our current concept of life). What is important is that an ontology yield scientifically fruitful generalizations for purposes of describing, explaining, predicting, manipulating, interpreting, and otherwise scientifically exploring the phenomena concerned. As the history of science underscores, there are often more advantageous ontologies waiting in the wings (so to speak). The best way to avoid premature commitment to an ontology is to entertain alternative ontologies as empirical evidence undermining those currently available accumulates (as opposed to sticking with them because they are intuitively appealing) and explore their differential capacities for supporting scientifically fruitful generalizations that maximize scope while minimizing exceptions. As discussed below, however, the tendency of most scientists is to do the opposite of what is being recommended.

4.4 The Threat Posed by Premature Commitment to Ontologies

The focus of this chapter is on the early stages of theory development. There is a reason for this. As adumbrated earlier and explained in the next chapter, our

---

[6] My use of the expression 'Goldilocks level of abstraction' is intended to remind astrobiologists of the Goldilocks zone of habitability. Just as the Goldilocks zone of habitability may contain more than one planet, so the Goldilocks level of abstraction may contain more than one scientifically fruitful ontology for supporting theoretically and empirically powerful generalizations. And just as one wants to focus attention on planets within the Goldilocks zone of habitability when searching for life that resembles ours, so one wants to search for scientifically fruitful, theoretical ontologies for a universal biology within the Goldilocks level of abstraction.

[7] Kyle Stanford's (2006) "problem of unconceived alternatives" is closely related, although he is concerned with whole theories as opposed to the ontologies that undergird their foundational principles.

current understanding of life is founded upon an unrepresentative subsample of a single example of life. Such a sample does not provide a secure foundation for generalizing about all life, wherever and whenever it may be found. The upshot is that the quest for a universal theory of life is still in its infancy. Nevertheless, the question arises as to whether what is being recommended for early stages of theory development applies to later stages, when what was once a highly successful scientific theory becomes overburdened by exceptions. It is beyond the scope of this chapter to investigate this interesting question. As this section discusses, however, at all stages of theory development, it is ontologies, as opposed to the principles in which they figure, that tend to be retained. This tendency is especially harmful in early stages of theory development and, I suspect, sometimes (but not always) detrimental at later stages. For in both cases it significantly constrains the range of potentially promising generalizations that can be formulated about a target phenomenal system.

The history of the concept of phlogiston – see, for example, Levere (2001) – provides a good illustration of the extent to which scientists are willing to go to preserve a favored ontology in the face of empirical exceptions. First proposed in the latter half of the seventeenth century to explain combustion, phlogiston was described as an imperceptible substance (lacking color, odor, taste, etc.) contained in combustible bodies and released when they burned. Substances that burn in air were characterized as rich in phlogiston and the fact that combustion quickly ceased in enclosed spaces was taken as evidence that air could absorb only a fixed amount of phlogiston. Experiments eventually revealed that some metals gain weight when they burn, which made it difficult to make sense of the claim that they had lost phlogiston. Most chemists did not, however, take this as evidence that the theory might be inadequate. Instead, they concluded that phlogiston either had negative weight or was lighter than air.[8] In other words, they focused on modifying the principles in which phlogiston figured rather than entertaining the possibility that the concept of phlogiston might be at fault. This almost certainly delayed the development of modern molecular theory.

Another illustration of the retention of a favored theoretical entity by modifying the generalizations in which it figures is provided by the concept of the luminiferous aether; for more discussion, see Whittaker (1910). Characterized as an insensible medium occupying every point in space, the aether was postulated to explain the mechanical propagation of light (later generalized to electromagnetic radiation) in "absolute" space; in classical physics, space is absolute (container-like) and all

---

[8] Hasok Chang (2014) argues that the phlogiston theory was prematurely abandoned. It is beyond the scope of this discussion to delve into the details of his argument. I will simply assume the standard view that Priestley and fellow travelers clung to it longer than was reasonable; even Kuhn (1970, 159), who was notoriously reluctant to characterize the theories abandoned in scientific revolutions as wrong, thought that this was the case.

motion is mechanical, which renders the idea of motion through empty space incomprehensible (a medium of transmission is required).[9] The failure of the Michelson–Morley experiments to detect the motion of Earth through the aether struck a serious blow to the concept. But instead of entertaining the possibility that the concept of aether might be the problem, new principles of aether behavior were proposed, such as that the aether remains attached to bodies as they move (aether drag) or that the length of a body is affected by the direction in which it moves through the aether (the contraction hypothesis). It was not until Einstein's theory of relativity (which rejects the Newtonian concept of absolute space, and hence a single universal frame of reference) that the concept of the luminiferous aether ceased to play a central role in physical theory.

The contemporary controversy over the viability of a universally applicable evolutionary concept of species exhibits the same pattern as the earlier debates over phlogiston and the luminiferous aether. Until fairly recently, it was widely assumed that the species concept, which lies at the core of Darwin's theory of evolution by natural selection, applies to all life, unicellular as well as multicellular (Hull 1999). But as Laura Franklin (2007) points out, classic versions of this concept in evolutionary biology, namely, the biological species concept (BSC) and phylogenetic species concepts (PSC), which are well suited to determining phylogenetic relationships among multicellular organisms (especially animals), break down when extended to prokaryotes (bacteria and archaea).

In classical Darwinian evolution, gene transfer occurs only during reproduction. Parents transfer genes vertically (via sexual reproduction) to their offspring. In contrast, some of the gene transfer that goes on in prokaryotes occurs independently of reproduction via a process known as horizontal gene transfer (HGT). A bacterium undergoing HGT acquires genes directly from another microorganism (which may not be closely related to it), as opposed to from its parent cell. When the bacterium divides, these newly acquired genes are transferred to its daughter cells. Depending upon how much HGT a bacterium (and its cell-based bacterial ancestors) has engaged in, it may have a multitude of different gene lineages; HGT is common among some bacteria and archaea. The upshot is that the same bacterium will qualify as belonging to different species, depending upon which genes are tracked, and many of the resultant species will not track the cell (organismal) lineage. Even worse, it may be difficult to pick out which gene lineage corresponds to a bacterium's cell lineage unless one can identify "core" genes for the cell lineage. It was once thought that certain ribosomal genes (16s RNA) were good candidates because they are rarely if ever subject to HGT. This supposition has

---

[9] Newton did not take a clear stand on this issue, in part, I suspect, because his theory of gravity does not require a medium for transmitting gravitational force. His successors, however, were strongly opposed to the notion of action at a distance.

been questioned in recent years, however (e.g., Tian et al. 2015). The upshot is that the concept of phylogenetic species for multicellular eukaryotes, which provides the foundation for biological systematics (e.g., determining phylogenetic relationships for purposes of constructing a "universal" tree of life), breaks down when applied to bacteria and archaea.

Franklin (2007) pins the problem on "... a neglect of bacterial biology in philosophical discussions of species and an almost exclusive focus on metazoans [animals]" (p. 70). The most popular response to the "bacterial species problem" among philosophers of biology is to embrace pluralism about species, that is, to deny that there is a single univocal concept of species that applies to all life. In keeping with the discussion in Section 4.3, however, there is another possibility: The very concept of a species might be the source of the difficulty. Perhaps (in a manner analogous to that of phlogiston and the luminiferous aether) the concept of species is scientifically unfavorable for purposes of characterizing and explaining evolutionary relationships *considered generally*. I will have more to say about this in Chapter 6.

As a final illustration of just how blind scientists can be to the role of an ontology in framing scientifically fruitful generalizations, consider the popular "big data" movement. The basic idea is to identify patterns and regularities among phenomena of interest by sorting through enormous quantities of data using sophisticated algorithms and statistical methods. In this way, it is claimed, one can by-pass the need for formulating and testing hypotheses and theories about the world. In a provocative essay on the topic, Chris Anderson (2008), a fan of the big data movement and a former editor of *Wired*, forecasts the "end of theory."

The big data movement tacitly assumes that "big" data encompasses *all* the data that there is about a phenomenon of interest. In fact, however, although voluminous, the data collected will comprise a biased subset of information. For, as Fulvio Mazzocchi (2013) points out, instruments and methods for collecting data are designed on the basis of scientific theories and background beliefs about the phenomena concerned. They cannot collect data that they were not designed to detect. An instrument designed for detecting amino acids cannot collect data on cosmic rays, for instance, and it will certainly not collect data on as yet unrecognized ways of classifying nature that may someday prove crucial for discovering the hidden unity underlying an ostensibly heterogeneous aspect of nature. The point is, despite Anderson's call for the end of theory, there is no such thing as an all-purpose, theory neutral, instrument or method for collecting everything that could be potentially known about an aspect of nature.

The theory laden instruments and methods used for acquiring voluminous data sets function in a manner similar to the replicas of the semantic conception. They collect only those classes of data that they were designed to detect, and hence implicitly carve the phenomena concerned into preconceived categories (an ontology) for searching for patterns and regularities. In denying that theoretical assumptions play any role in their investigations, proponents of big data fail to appreciate this. As a result, the ontologies implicitly underlying their research become even more deeply entrenched (literally unrecognized as such) than in research that is explicitly theory driven. Failure to identify truly general patterns and regularities is thus very unlikely to be interpreted as signifying a potential problem with the theoretical lenses (instruments and methods) through which they are exploring the world. Instead, it will be taken as reflecting the objective (theory neutral) character of the phenomena concerned. In other words, the big data movement is more likely to hinder than advance our scientific understanding of nature.[10]

The tendency to retain (and even be blind to) ontologies and modify generalizations when faced with ostensibly disconfirming empirical evidence is not surprising. As Ian Hacking (1983) points out, most scientists are realists when it comes to the basic entities postulated by their theories – or in Nancy Cartwright's (1983, e.g., p. 6) terminology, the theoretical "causes" of observable natural phenomena – but not about the theoretical generalizations in which they figure. An intriguing illustration is recent experimental work on whether Newton's inverse-square law of gravity breaks down under certain conditions present right here on Earth (e.g., Eckhardt et al. 1988 and Ander et al. 1989).[11] In contrast, physicists are much less willing to consider giving up the concept of mass, a key theoretical concept in Newton's physics that is retained in contemporary physics. Significantly, however, the *laws* in which mass enters have changed since Newton's theory. In Newton's theory mass is conserved. It is not conserved in Einstein's special theory of relativity, where mass is treated as interconvertible with energy (as in $e = mc^2$), and mass-energy is viewed as jointly conserved.

In this context, I hasten to remind the reader that the focus of this chapter is on the early development of scientific theories. Our concern is with the quest for a universal theory of life which, as adumbrated above and discussed in much greater detail in the next two chapters, is still in its infancy. The status of mature scientific theories, such as Newton's and Einstein's, is different. It makes good sense to stick with core theoretical concepts (e.g., mass) that have proved their mettle by supporting generalizations that, despite facing an increasing number of exceptions,

---

[10] It is important to distinguish (what I have dubbed) the big data movement from the use of sophisticated algorithms and statistical methods to sift through massive quantities of data searching for patterns and regularities. There is no question that one can discover interesting patterns and regularities by means of such techniques. The mistake is in thinking that these findings (or lack thereof) do not presuppose a theoretical ontology for structuring the phenomena being investigated – that they represent unvarnished objective facts about nature.

[11] I hasten to add that I take no stand on this issue. I am merely illustrating a point: Scientists are far more resistant to responding to theories overburdened with exceptions by adjusting ontological commitments than they are to amending them by modifying basic principles.

have historically proven highly successful for purposes of prediction, explanation, etc. In such cases, revising generalizations or reinterpreting (versus jettisoning) key concepts makes more sense than starting over with brand new theoretical concepts in the hope of discovering more fruitful unifying principles. Still, there is always the risk that the problem lies in the concepts, as opposed to generalizations in which they figure. It is beyond the scope of this chapter, however, to address the interesting question of when a mature scientific theory becomes so encumbered with seemingly intractable exceptions that it becomes prudent to entertain alternative theoretical frameworks for reasoning about the phenomena concerned.

4.5 The Monist (Versus Pluralist) Stance

Earlier chapters spoke somewhat disparagingly of scientific pluralism. Yet the sections above seem to be advocating pluralism about ontologies. What is going on?

A brief survey of the better known versions of scientific pluralism is helpful for purposes of orientation. Modest versions treat the world as a heterogeneous patchwork of natural phenomena which cannot be unified under a theory of everything. As Kellert et al. (2006, xii) point out, in their widely cited introduction to *Scientific Pluralism*, however, modest pluralism does not reject the notion that there is (at least in principle) a single best theory for each patchwork. More radical versions of scientific pluralism, on the other hand, maintain that there is no unique best theory for even a single patchwork of nature. John Dupré (1993), who embraces the most radical form, which he calls "promiscuous realism," contends that there are an indefinite number of ways of divvying up the world (and its parts) into categories for purposes of scientific practice and argues that there is no answer to the question of which is more correct. While not as permissive as Dupré, insofar as he does not think that anything goes, Hasok Chang (2014) nonetheless contends that it is a mistake to suppose that there is a unique best theory for most aspects of nature. Appealing to the infamous Duhem problem,[12] he argues that scientists should not jettison old theories for new theories but instead continue to develop both alongside one another. Chang's poster child is phlogiston theory, which he argues was prematurely abandoned in the nineteenth century; he sketches how it might be resurrected today as a viable competitor to molecular theory. The fact that new theories are often incompatible with their predecessors (in that they say

---

[12] Pierre Duhem was a nineteenth century French physicist, historian, and philosopher who argued that no scientific hypothesis can be tested in isolation from assumptions about background conditions, which include assumptions about equipment (e.g., that it is working properly) and contingencies (the absence or presence of interfering factors); these assumptions are known as auxiliary assumptions/hypotheses. The upshot is the Duhem problem: Disconfirming empirical evidence can always be plausibly handled by rejecting an auxiliary assumption instead of the target hypothesis.

conflicting things about the very same aspects of nature) does not bother Chang. In his words, ". . . we should pursue *all* systems of knowledge that can provide us an informative contact with reality; if there are mutually incompatible paradigms, we should retain all of them at once" (p. xix).

Both modest and radical pluralism rest on metaphysical suppositions about the world, namely, that it is *at least* intrinsically disorderly as a whole, and depending upon the version of pluralism, some or all of its parts are also intrinsically heterogeneous. Scientific monism also rests upon metaphysical suppositions about the world insofar as it holds that the world could (at least in principle) be accounted for scientifically in terms of a comprehensive set of fundamental principles; it is committed to there being an underlying, as yet undiscovered, unity to the manifest heterogeneity of nature. As with pluralism, there are different versions of monism. Most do not claim that humans, with their limited intellects, could actually acquire such a theory even though they endorse the quest for theories that ever more closely approximate it. Furthermore, not all versions of monism are committed to the possibility of a comprehensive and complete (reductionist) theory of everything (e.g., Craver 2005; Schaffner 2007). One could be a monist about biological phenomena, for instance, without endorsing the view that biological phenomena are fully explicable in terms of elementary particle physics. As the reader may surmise, the distinction between modest pluralism and modest (nonreductive) monism is as thin as gossamer from a metaphysical perspective.

This book does not take a stand on the metaphysical issue of what underlies the pervasive empirical manifestations of disunity that (as Cartwright, Wimsatt, Mitchell, and so many others, have noted) afflict all of our sciences: Is it the immense complexity of nature, which makes it difficult, if not impossible, for a limited mind like ours to comprehend, let alone formulate, such a theory (for the world as a whole or, barring that, some aspects of it)? Or is the world (or at least many aspects of it) truly pluralistic in the metaphysical sense that no such theory is even possible? These questions, while important, are beyond the scope of this book. My concern is with an epistemological issue about a particular aspect of nature, namely, the phenomena of life: Is (as many pluralists contend) the *program* of universal biology fundamentally mistaken?

Kellert and colleagues (2006, xii–xxvii) advocate a "pluralist stance," by which they mean ". . . a commitment to avoid reliance on monist assumptions in interpretation or evaluation coupled with an openness to the ineliminability of multiplicity in some scientific concepts" (p. xiii). The authors claim to be agnostic as to whether pluralism or monism is true of nature: ". . . we do not assume that the natural world cannot, in principle, be completely explained by a single tidy account; we believe that whether it can be so explained is an open empirical question" (p. x). I agree with them that whether pluralism or monism is true, either of nature as a whole or of particular aspects of it, is an open question. Where I disagree is with their contentions that (i) the fact that none of our current scientific theories are complete provides compelling empirical evidence that monism is false and (ii) a pluralist stance towards investigating the world scientifically is more beneficial to science than a monist stance.

A pluralist stance towards the world is commonly defended on the grounds that our most successful scientific theories are burdened with exceptions. As Chang puts it, ". . . it has always been the case that even our best scientific theory fails to cover all phenomena" (2014, 272). Similarly, Kellert and colleagues counsel, "[a] pluralist stance keeps in the forefront the fact that scientific inquiry typically represents some aspects of the world well at the cost of obscuring, or perhaps even distorting other aspects" (Kellert et al. 2006, xv). Chang, Kellert, Longino, and Waters diagnose our failure to come up with truly exceptionless universal theories of natural phenomena as *prima facie* evidence that monism is false.

As argued in Section 4.3, however, there is another, equally plausible, interpretation: Theoretical unification requires idealization, that is, abstracting away from the messy contingent details of a *particular* phenomenal system in order to lay bare general patterns holding for phenomena *of that kind*. The upshot is that even supposing that monism is true, it is unlikely that any of the theories that we come up with will *manifest* as exceptionless. As discussed earlier, it is difficult to determine whether an alleged exception to a scientific theory is the product of poorly understood contingencies lying outside of the scope of the theory, a defective theoretical principle(s) in need of revision or replacement, or the intrinsic heterogeneity of the phenomena of interest. Accordingly, the fact that even our best scientific theories face "exceptions" does not provide a good argument against monism. For even supposing that there is an underlying unity to all the troubling manifestations of disunity exhibited by nature, it is naïve to suppose that science will someday provide us with theories that are unambiguously exceptionless. Most importantly for our purposes, this is especially true of theories in the special sciences, such as biology, whose exceptions are commonly touted as providing the best evidence for pluralism. On the other hand, it is not naïve to seek theories that do an increasingly better job of maximizing scope while minimizing exceptions than those that we currently possess. I dub this perspective on scientific practice the "monist stance."

Contrary to Kellert and colleagues' (2006, xv) and Chang's (2014, 259) stereotype of monism, the monist *stance* does not claim that there is one true theory and that it is the job of scientists to figure out which one it is. It does not even claim that there is a single "best" theory for a given aspect of nature. One cannot eliminate the possibility of alternative ways of subdividing a domain of phenomena into categories that are equally good at supporting scientifically fruitful generalizations. What the monist stance stresses is the importance of replacing current scientific theories by theories that are even more comprehensive and complete. The strategy recommended for doing this in the early stages of theory development, which is the focus of this book, is to explore the capacities of alternative ontologies (rival sets of theoretical concepts) for supporting generalizations that achieve a better balance between maximizing scope and minimizing exceptions (while satisfying other important scientific virtues).

Despite acknowledging the possibility of equally "good" but different scientific theories for a given domain of natural phenomena, the monist stance rejects Chang's recommendation that scientists simultaneously develop rival theories. The problem with Chang's approach is that it reduces the pressure to seek new and better scientific theories. There are only so many competing theories for an aspect of nature that scientists can simultaneously pursue. Proliferating without eliminating theories will almost certainly slow down the development of more comprehensive and inclusive theories. Indeed, Chang's proposal is reminiscent of what Kuhn calls "pre-paradigmatic science" where researchers align themselves with rival schools of thought in a tribal fashion and rarely if ever abandon the theories to which they are committed; in the face of ostensibly disconfirming evidence, they persist in revising their theories by fiddling with auxiliary assumptions (about the experimental situation, background conditions, etc.), which as Duhem pointed out, is always logically permissible.

This brings us to the allegation that a pluralist stance is more beneficial to science than a monist stance. Instead of directly addressing arguments that have been advanced in support of this claim, many of which appeal to allegedly discipline specific variations in epistemic values and aims, I want to turn the table and provide some compelling reasons for preferring a monist stance. First, regardless of what pluralists claim, the discovery of underlying unifying principles of nature is perhaps the oldest and most central goal of science. Such principles are traditionally held up as providing our deepest and most satisfying understanding of nature. I am not claiming that this is the only aim of science but it is certainly a central one in the minds of many scientists. In stressing other aims over unification, pluralists disregard this important fact.

Second, and more importantly, the pluralist stance advocated by Kellert, Longino, and Waters is not as metaphysically innocent as they contend. Someone who takes a pluralist stance towards a domain of natural phenomena is unlikely to recognize unity when they encounter it. They will not be anticipating unity, let alone searching for it. Instead, they will be expecting heterogeneity, and this is what they will find. For as discussed earlier, nature does not come pre-structured into theoretical categories supporting scientifically fruitful general principles. It takes both theoretical and empirical work to discover unity in the manifestation of disunity. On a pluralist stance, there is little incentive to do this work.

In contrast, while remaining open to the possibility that a domain of ostensibly heterogeneous phenomena lacks unity, the monist stance nonetheless demands that researchers persist in seeking it. The advantage of the monist over the pluralist stance is that it is easier to recognize disunity in a search for unity than it is to recognize unity in the face of an expectation of disunity. Disunity will manifest as exceptions to the conjectured principles whose scope and robustness are being hopefully investigated. Instead of reconciling themselves to these pesky exceptions, researchers taking a monist stance will search for principles that achieve a better balance between scope and exceptions. Even supposing (as seems likely) that they never hit upon a final "best" theory, their efforts will generate increasingly more unifying theories until (perhaps) a point is reached at which the possibilities for maximizing scope and minimizing exceptions have been pretty much exhausted. Significantly, there is no evidence that we have reached such a point yet for any aspect of nature, and this is especially true of biology.

4.6 Concluding Thoughts

The aim of this chapter is not to provide a comprehensive account of the development of successful scientific theories, which itself would be a book length project. Instead, the focus has been on identifying factors that hinder the development of such theories with an eye to facilitating the construction of a universal theory of life. As Section 4.2 explained, in the context of illustrations from the history of science, the development of a successful scientific theory critically depends upon finding a propitious set of foundational theoretical concepts (an ontology) for formulating scientifically fruitful general principles. Select the wrong ontology and such generalizations will not be forthcoming. The identification of matter with bulk (occupied volume), impenetrability, and weight by Newton's predecessors frustrated the discovery of universal laws of motion. Newton succeeded in formulating such principles because he availed himself of a more promising foundational theoretical concept, namely, that of inertia.

Given the challenge, discussed in Section 4.3, of striking the right (Goldilocks) level of abstraction for theorizing about a domain of natural phenomena, and the fact, discussed in Section 4.4, that scientists tend to settle fairly quickly on ontologies and retain them in the face of predictive and explanatory failure of the generalizations in which they figure, it is especially important to avoid allegiance to an ontology in the early stages of theory development. For as the history of science reveals, there is no guarantee that an intuitively pleasing way of conceptually carving an aspect of nature into theoretical categories will support generalizations falling within the Goldilocks zone of abstraction. This worry is especially acute in cases, such as life, where our experience is clearly very limited; as the next chapter (Chapter 5) discusses, biochemists have identified ways in which life could differ from familiar life at the molecular level and they do not know how different it could be. The tendency is to formulate generalizations about life that are extremely abstract. Popa (2004), for instance, opines that a comprehensive account of life must "... identify properties that are independent of its physical nature" (p. 3), a view that is endorsed by ALife researchers defending purely informational forms of life (Chapter 7). Astrobiologists who design life detection strategies based closely on the biochemistry of familiar life (Chapter 8), on the other hand, implicitly endorse too concrete an account of life. The Goldilocks level of abstraction for theorizing about life almost certainly lies somewhere in between these extremes. The question is where?

The most promising strategy for increasing the chances of finding a propitious set of theoretical concepts for theorizing about a domain of phenomena is to formulate and explore rival ontologies (avoiding extremes of abstractness and concreteness) for their differential capacities for supporting generalizations that maximize scope while minimizing exceptions. In a case such as life, where scientists are dealing with a single, possibly unrepresentative example, this process should continue as newer and more unusual forms of life are discovered. For as Chapter 5 discusses, our understanding of familiar Earth life is quite limited. Complex multicellular eukaryotes, which still ground much of biological theorizing, are rare and exotic latecomers to our planet. Bacteria and archaea are by far the oldest and most representative form of Earth life, and while our understanding of the microbial world is still fairly rudimentary, it is increasing rapidly. There is a sense in which the microbial world is already providing us with novel forms of life, and the prospects for discovering even more unusual Earth microorganisms, perhaps even a shadow biosphere (Chapter 9), are good. One does not want to hamper the development of a scientifically more fruitful theory of life by insisting on fitting the microbial world into a theoretical framework designed on the basis of what we now know are outliers of familiar life.

This brings us back to an issue raised in Chapter 1. As accentuated in popular (metabolic and genetic) definitions of life, Aristotelian concepts (of self-nutrition and self-reproduction) still dominate scientific thought about the nature and origin of life. Could a defective neo-Aristotelian ontology for life be responsible for the failure of biologists to identify truly general principles even for familiar life? Indeed, there is an uncanny resemblance between the neo-Aristotelian concept of impetus (an internal capacity of an inanimate object to move itself), which dominated physics before Newton, and contemporary concepts of metabolism and reproduction (with their implicit appeals to self-causation). Is this the real reason why (to paraphrase Kant) we have never had a Newton of biology?[13]

There are of course other potential explanations for the failure of biologists to identify truly general principles for life. Perhaps familiar Earth life is unrepresentative of life, and in the absence of additional examples, we have been unable to distinguish contingent from fundamental characteristics of life, which only serves to underscore the importance of exploring alternative ontologies at this stage of biological thought. In any case, however, that biologists have yet to come up with truly general principles for biological phenomena does not provide strong support for the widespread pluralist view that a universal theory of biology is impossible (because life is not a natural kind). For as suggested in Section 4.5, the manifestation of heterogeneity among known biological phenomena could be little more than an artifact generated by a flawed theoretical framework founded upon an unrepresentative subsample of life.

What is really needed to answer the age-old question 'what is life?' is a scientific theory founded upon additional examples of life. Chapters 7, 8, and 9 explore three potential strategies for acquiring the needed examples in the absence of either a definition or universal theory of life. In lieu of discovering additional examples of life – which may not happen for a long time, and moreover is not guaranteed – Chapter 6 suggests another tactic: Instead of trying to fit the known microbial world into a theoretical framework founded upon complex multicellular eukaryotes, hunt for new ontologies for reasoning scientifically about life among the microbes. For as Chapter 5 explains, archaea and bacteria are the oldest and most representative form of life on Earth, and hence far more likely than multicellular eukaryotes to yield a set of core theoretical concepts capable of supporting empirically fruitful, truly general principles about Earth life (if not life in general).

---

[13] As discussed in Section 1.3, Kant argued that there would never be a Newton of biology because the idea of teleology could not be accommodated within a Newtonian framework.

5

Challenges for a Universal Theory of Life

5.1 Overview

The most significant challenge facing the pursuit of a universal theory of life is the infamous "$N = 1$ problem." In the late twentieth century biologists made an astonishing discovery. Life as we know it on Earth today descends from a last universal common ancestor (LUCA), and hence represents a *single* example of life. Logically speaking, one cannot safely generalize to all of life, wherever and whenever it may be found, on the basis of a single example. As Section 5.2 explains, the $N = 1$ problem of biology is not just a pernickety logical point. There are compelling scientific reasons for worrying that our sample of one may be *unrepresentative* of life. Biochemists and molecular biologists have established that life could differ from familiar Earth life in significant ways at the molecular and biochemical levels. In addition, astrobiologists have explored how the basic functions of familiar life (metabolism and genetic-based reproduction) might be realized by molecular compounds based on elements other than carbon under chemical and physical conditions differing from those thought to have been present on early Earth.

Section 5.3 addresses yet another, little discussed, facet of the $N = 1$ problem of biology: Most scientific theorizing about life uses concepts and principles founded upon an unrepresentative subsample of Earth life. By the end of the twentieth century, it had become clear that microbes are the most representative form of life on Earth. Bacteria and archaea are at least 3 billion years older than complex multicellular eukaryotes. They are also environmentally tougher and far more diverse metabolically than eukaryotes. Complex multicellular eukaryotes, which include plants and animals, are now recognized to be exotic, highly specialized, latecomers to our planet. The upshot is that they are not representative of Earth life. In essence, modern biology is built on an unrepresentative subsample of a single example of life.

The status of complex multicellular eukaryotes as outliers has consequences for the program of universal biology. If one is going to pursue a truly general theory of life in the context of our current epistemic limitations – that is, in the absence of examples of life descended from an alternative abiogenesis – then surely the most promising approach is to seek an ontology (core theoretical concepts) for generalizing about life from among the most representative form of life available to us, namely, Earth microbes. Indeed, what we are learning about the microbial world is challenging some of the most venerable concepts and tenets of modern biology, which, as Chapter 6 discusses, are founded upon complex multicellular eukaryotes. In a very real sense, the microbial world is already supplying us with examples of life as we don't know it.

Another obstacle to developing a universal theory of life is the expectation that such a theory will tell us everything there is to know about life. The job of a universal theory is to capture the unity behind the diversity within a domain of natural phenomena. Capturing the unity behind the diversity of a domain of natural phenomena does not, as Section 5.4 explains, mean being able to explain everything there is to know about the phenomena concerned. In truth, as Section 4.3 explained, no scientific theory is truly exceptionless. Individual (actual) occurrences are invariably subject to contingencies lying outside the scope of even the most general theory. Models of the origin(s) of life provide especially salient illustrations of the tension between individuality and generality. The RNA World and SM (Small Molecule) World are fashioned closely on "definitions" of life, which in virtue of supplying necessary and sufficient conditions for life are universal. As Section 5.4.1 explains, both models are riddled with puzzles about contingencies. This is not surprising because an origin of life, whether it occurs once or multiple times, is an individual, undoubtedly complex, and, most importantly, highly contingent, event. No two origins of life will be exactly the same and some of them may be notably different. If (as some hopeful astrobiologists speculate) there is life on Saturn's moon Titan, the geochemical conditions of its origin are likely to be very different from those in which familiar life arose on Earth; Titan has no liquid water and lakes of methane.

Viewed in this light, the tendency of scientists and philosophers to base theories of the origin(s) of life closely on sweeping generalizations about the nature of life is perplexing: It assumes that one can disregard contingencies and (in essence) extract a recipe for synthesizing life from an account of its nature. As Section 5.4.2 explains, this supposition is false for other natural kinds, and there is little reason for supposing that life is somehow an exception. Failure to disentangle contingency-ridden models of the origin of life from sweeping generalizations about the nature of life (whether presented in the form of definitions or theories) can thwart the development of scientifically compelling models of the origin of life. It can also prematurely cast doubt on a fledgling theory of life by demanding more of it than any universal theory could deliver.

5.2 The Magnitude of the $N = 1$ Problem of Biology

This section explains the extent of the threat posed by the $N = 1$ problem of biology to scientific and philosophical reasoning about the nature of life. As discussed below, there are compelling empirical and theoretical reasons for worrying that familiar Earth life not only provides us with a single example of life but also that it may not be representative of life.

From bacteria, slime molds, rotifers, jellyfish, mushrooms, grasshoppers, snakes, birds, redwood trees, and elephants, life as we know it on Earth today is remarkable for its morphological diversity. On the basis of morphology alone, one might suspect that all the possibilities for life are represented right here on Earth. Studies of the molecular and biochemical basis of familiar life indicate otherwise, however. Familiar life uses the same carbon-containing macromolecules (proteins and nucleic acids) and molecular architecture (ribosomes) to realize the biological functions traditionally held up as essential to life (metabolism and genetic-based reproduction). Proteins supply the bulk of the structural material for building organismal bodies as well as the catalytic material (enzymes) for powering them. The genetic system uses nucleic acids to store and process hereditary information: DNA (deoxyribonucleic acid) stores hereditary information and RNA (ribonucleic acid) orchestrates its translation into proteins. The process of coordinating these life conferring, biological functions – of translating hereditary information stored in DNA into proteins for use in growth, maintenance, repair, and reproduction – is handled by ribosomes, which are highly complex, minuscule, molecular machines, composed of RNA and protein, that are found in very large numbers in every cell. A single *E. coli* cell contains around 15,000 ribosomes. Some mammalian cells contain as many as 10 million.

The molecular and biochemical similarities among known Earth organisms go even deeper into the composition and structure of their proteins, and it is here that biochemists have identified specific ways in which life could be different. Proteins typically consist of 50–1000 amino acids joined together by peptide bonds into long chains or polymers (Figure 5.1). Amino acids are fairly complex molecules consisting of two functional groups, a basic amino group ($-NH_2$) and an acidic carboxyl group ($-COOH$).[1] Known life on Earth uses the same 20 (directly genetically encoded) amino acids to construct its proteins even though this represents a small subset of the more than 100 amino acids found in the natural environment. A protein's biological functionality depends upon its capacity to fold into a complex three-dimensional structure, which in turn depends upon (but is

---

[1] An amino group has electrons available for binding with another atom, making it a base, and the carboxyl group has a proton available for binding with another atom, making it an acid. [Note: Technically speaking, at neutral pH almost all of the amino acid is in ionic form (COO^- and NH_3^+) but the reaction itself occurs between the neutral forms ($COOH$ and NH_2).]

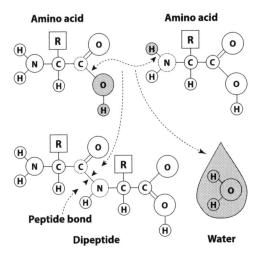

Figure 5.1 Amino acids differ only in the molecular unit represented by R above; they all share a basic amino group (–NH$_2$) [top images on the left] and a basic acidic group (–COOH) [top images on the right]. The bottom image shows a peptide bond between the two amino acids, which released a water molecule. [Note: Technically speaking, at neutral pH almost all of the amino acid is in ionic form (COO$^-$ and NH$_3^+$) but, as indicated in the figure, the reaction itself occurs between the neutral forms (COOH and NH$_2$).]

not fully determined by) characteristics of its constituent amino acids (Alberts et al. 2002). There must be a sufficient number of small, large, hydrophobic (water repelling), hydrophilic (absorbing or dissolving in water), and charged amino acids, and they must occur in the correct sequence; modifying the order changes the functionality and hence identity of a protein. But while this restricts which collections of amino acids are capable of synthesizing a sufficiently diverse collection of proteins for embodying and powering an organism, it by no means eliminates alternatives to the standard collection used by familiar life.

Biologically functional proteins might differ from those used by familiar Earth life in yet another way. Amino acids have the geometrical property of handedness or, more technically, chirality in virtue of their chemical bonds being asymmetrically arranged around a carbon atom (Meierhenrich 2008). One of these chiral arrangements is known as "L" (levo) and the other as "D" (dextro). The genetically encoded proteins of familiar life are synthesized exclusively from L-amino acids. Because racemic mixtures of L-amino acids and D-amino acids do not build good protein structure, it is hardly surprising that they are not synthesized from heterochiral mixtures. The question is why L-amino acids instead of D-amino acids?

While the alternative molecular possibilities canvassed above are theoretical, based on our current understanding of biochemistry, there are compelling empirical grounds for taking them seriously. Alternative proteins that fold correctly and would be functional in the right organismal environments have been synthesized in the laboratory from standard D-amino acids, nonstandard L-amino acids, and nonstandard D-amino acids (e.g., Benner 1994). Moreover, there are a variety of potential abiotic sources for standard and nonstandard amino acids of both chiralities, including meteorites (Cronin and Pizzarello 1983; Pizzarello et al. 2006), photochemical reactions in atmospheric aerosols (Dobson et al. 2000), and chemical reactions in oceanic hydrothermal vents (Amend and Shock 1998; Marshall 1994; Martin et al. 2008; Russell and Hall 1997). As an illustration, over seventy amino acids, only eight of which are used by familiar life, have been identified in meteorites (Schmitt-Kopplin et al. 2010); studies suggest an excess of L-amino acids but D-amino acids are also present (see also Pizzarello and Shock 2010). These empirical findings suggest that the standard set of L-amino acids used in the proteins of familiar life are the product of chemical and physical contingencies at the time of the origin of life on the early Earth.

Similar considerations apply to the molecular composition of nucleic acids, which make up the hereditary material of all known life on Earth. Nucleic acids consist of long polymers of monomeric nucleotides, which are more complex than the amino acids comprising proteins. Nucleotides are built from three subcomponents, a negatively charged phosphate unit, a sugar unit (deoxyribose in DNA and ribose in RNA), and one of five standard (nucleo)bases: adenine (A), guanine (G), cystosine (C), thymine (T), and uracil (U). They are chained together into nucleic acid polymers by alternating sugar (S) and phosphate (P) units, which form a multiply charged backbone. The bases, which encode genetic information, are attached to the sugar units. The famous double helix of DNA consists of two strands (chains of nucleotides) held together by hydrogen bonds between their respective bases (Figure 5.2). The four bases of DNA are A, C, G, and T, and they pair with each other in a complementary pattern, namely, C to G and A to T; RNA substitutes U for T.[2]

The biological function of nucleic acids is to store hereditary information and supervise its translation into proteins, which are used to build, maintain, and power organisms. Hereditary information is stored on a single (the coding) strand of DNA by the bases using a triplet code. Each triplet of bases (codon) corresponds to an amino acid or a "start" or "stop" signal for constructing a protein. The translation of hereditary information into proteins involves a multistage physicochemical process in which information carried by DNA is transcribed onto messenger RNA (mRNA), transported to a ribosome, and translated into protein with the help of transfer RNA (tRNA) (Figure 5.3).

---

[2] Unlike DNA, RNA is typically single stranded but contains secondary structures, folded regions where complementary bases on the same strand bind together.

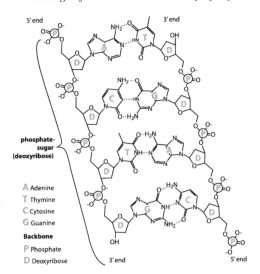

Figure 5.2 A portion of a DNA molecule, showing the two strands (chains of nucleotides), with their negatively charged phosphate–sugar (deoxyribose) backbones [far left and far right] and the hydrogen bonds [dotted lines] between their respective (nucleo)bases [see, e.g., adenine and thymine on the top tier] that hold the two strands together.

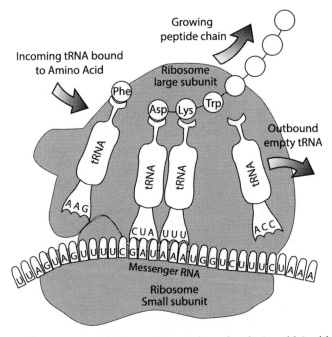

Figure 5.3 A ribosome assembles a small protein molecule (peptide) with the help of tRNA, which supplies particular amino acids (e.g., lysine) in the order specified by mRNA (triplets of bases, e.g., AAA, transcribed from DNA).

RNA Codon table

1st position	2nd position →	U	C	A	G	3rd position
U		Phe	Ser	Tyr	Cys	U
U		Phe	Ser	Tyr	Cys	C
U		Leu	Ser	*stop*	*stop*	A
U		Leu	Ser	*stop*	Trp	G
C		Leu	Pro	His	Arg	U
C		Leu	Pro	His	Arg	C
C		Leu	Pro	Gin	Arg	A
C		Leu	Pro	Gin	Arg	G
A		Ile	Thr	Asn	Ser	U
A		Ile	Thr	Asn	Ser	C
A		Ile	Thr	Lys	Arg	A
A		Met	Thr	Lys	Arg	G
G		Val	Ala	Asp	Gly	U
G		Val	Ala	Asp	Gly	C
G		Val	Ala	Glu	Gly	A
G		Val	Ala	Glu	Gly	G

Amino Acids

Figure 5.4 The genetic code. Note: There are 20 amino acids. Also, the code is redundant. Some amino acids (e.g., lysine) are encoded by more than one codon (AAA and AAG). But no codon encodes more than one amino acid.

Biochemists have explored molecular variations in all three of the subunits of nucleotides. The five bases used by standard nucleic acids are not the only chemically viable possibilities for encoding genetic information. At least eight additional bases capable of forming duplex DNA have been identified and tested in the laboratory (Benner 2004; Benner et al. 2004).

The genetic code (Figure 5.4) also seems somewhat arbitrary. While no codon is paired with more than one amino acid, many amino acids are paired with more than one codon, providing redundancy and, most importantly for our purposes, indicating that there are multiple molecular possibilities for encoding amino acids. Indeed, researchers have artificially modified the genetic code by pairing nonstandard amino acids (i.e., amino acids not encoded by any natural organism on Earth) with redundant codons in bacteria and yeast cells (Xie and Schultz 2005). Furthermore, the use of a triplet coding scheme is contingent upon the number of bases available for encoding genetic information and the number of amino acids used to construct proteins. As is sometimes noted, a triplet coding scheme is the most efficient for 4 bases and 20 amino acids. But there is little reason to suppose that an alternative form of life could not use different numbers of bases or amino acids, in which case, a doublet, quadruplet or even quintuplet coding scheme might be more efficient than a triplet coding scheme. Some synthetic biologists are exploring possibilities along these lines (Baross et al. 2007, Sec. 5.1). The development of an artificial genetic code consisting of two alternative bases that can be incorporated into DNA along with the standard four (Yang et al. 2011) is especially noteworthy.

Biochemists have also investigated alternatives to the sugar–phosphate backbone of nucleic acids. Biologically important features of the backbone include the ability to form an extended, redundant, stable structure that is insensitive to the precise order of the bases, which permits DNA and RNA to encode and preserve an enormous amount of hereditary information through variations in base sequence. To be capable of evolving a genetic system, a nucleic acid analog must also produce low frequency errors (as a source of heritable variation) that can be replicated with high fidelity. It is difficult to construct nucleic acid analogs having all of these features (Benner and Hutter 2002; Benner et al. 2004). As an illustration, peptide nucleic acids (PNA), which replace the multiply charged sugar–phosphate backbone of standard nucleic acids with an uncharged peptide-like backbone built from *N*-(2-aminoethyl)glycine units is able to form a double helix by means of hydrogen bonds between the standard complementary bases (Nielsen and Egholm 1999). But because the repeating backbone charge supplied by the phosphate units is missing, PNA lacks the stability of DNA. Small changes in base sequence can dramatically change its molecular behavior, sometimes even preventing the formation of a duplex. PNA thus lacks both the stability and insensitivity to base order that make RNA and DNA such good storage mechanisms for hereditary information.

In light of this problem, recent efforts have focused on building an alternative backbone by retaining the charged phosphate units while replacing the sugar unit. A breakthrough was recently achieved using XNA nucleic acids, which replace the standard sugar units with alternative carbohydrates (while retaining the bases and charged phosphate units of standard nucleic acids), to construct an alternative genetic system having the capacity not only to store but also to retrieve, propagate, and evolve heritable information (Pinheiro et al. 2012). As the authors note, their work with XNA suggests that "... heredity and evolution, two hallmarks of life, are not limited to DNA and RNA" (p. 341). In principle, one could construct even more exotic genetic systems from XNA molecules by fiddling with the identity or number of bases, for example, adding the two new bases mentioned earlier to the standard four to create a six-base system.

In sum, despite its astonishing morphological diversity, all known life on Earth is remarkably similar at the molecular level. Moreover, biochemists have identified a variety of highly plausible ways in which the basic biomolecules of familiar life could be different without compromising their biofunctionality. As a consequence, there is widespread agreement among biochemists that the genetic code and some of the molecular components of proteins and nucleic acids represent "frozen accidents," that is, physical and chemical contingencies present at the time that life emerged on the early Earth (e.g., Benner et al. 2004; Smith and Morowitz 2004). These contingent molecular similarities, coupled with studies of highly conserved genomic sequences from unicellular and multicellular organisms, provide strong support for the conjecture that familiar life on Earth descends from a last universal common ancestor, and hence represents a sample of one.

Generalizing on the basis of a sample of one is always risky unless one has good reasons for believing that the sample is representative. There are compelling empirical and theoretical reasons, however, for suspecting that familiar life on Earth is *not* representative. As just discussed, chemists have identified specific ways in which the core biomolecules of life could be at least modestly different. The question is how different: Are proteins and nucleic acids the only polymers based on carbon that are capable of performing core biological functions? Could polymers based on elements other than carbon perform the basic functions of life? As discussed below, given the current state of our knowledge, it is very difficult to constrain the possibilities.

Compared to a typical molecule found in nature, DNA seems highly improbable. It is huge and extremely complex, built from repeating units composed of large numbers of a variety of chemical elements (carbon, hydrogen, oxygen, phosphorus, and nitrogen). Even today no one knows how to synthesize it or the monomeric units – nucleotides, or for that matter, nucleosides (sugar–base compounds) – that comprise it from molecular precursors by means of naturally occurring abiotic processes (e.g., Baross et al. 2007, Sec. 5.4; Shapiro 2000). The point is DNA is not the sort of molecule that could be inferred from purely theoretical, chemical considerations or accidentally concocted in a laboratory in an exploratory experiment.

Similar considerations apply to evaluating the prospects for noncarbon-based life. Because it is the only element other than carbon known to form extended polymers, silicon is the most popular candidate. Chemists have not, however, been able to synthesize silicon polymers in the laboratory having anything near the complexity and versatility thought to be required of molecules capable of encoding and evolving hereditary information or building and powering organismal bodies. Nevertheless, the jury is still out. Schulze-Makuch and Irwin (2006) argue that silicon may be capable of forming more promising molecular structures under conditions quite different from those found on Earth today, namely, where there is little liquid water, little oxygen, and very low temperatures. Similar considerations apply to other chemical elements whose capacities for forming complex molecular structures of the right sort, under physicochemical conditions very different from those on Earth, have yet to be adequately explored. Viewed from this perspective, extraterrestrial chemists from an appropriately un-Earth-like environment, such as that found on the surface of Saturn's moon Titan or perhaps even the atmospheres of Venus (e.g., Grinspoon 1997; Schulze-Makuch and Irwin 2006) or Jupiter (Benner et al. 2004), might be just as convinced, on the basis of *their* theoretical understanding and laboratory experiences, that while carbon is capable of forming simple, extended molecular structures, such as graphite or diamond, it is nonetheless incapable of forming functional biomolecules.

As chemists Powner and Sutherland (2011) point out in a provocative paper, laboratory studies of the origins of life are especially problematic. They typically focus on linear, multistep reaction sequences involving simple mixtures of chemicals under constant conditions. Chemical mixtures found in natural environments capable of producing and evolving life (such as the early Earth, or perhaps Titan or even Jupiter), however, are far more complex with many interconnected reactions going on under fluctuating geochemical and geophysical conditions. The authors conjecture that such considerations may shed light on, for example, the failure of chemists to explain how ribonucleotides – RNA monomers – arose on the early Earth under abiotic conditions.

To wrap up, in addition to the traditional logical concerns about generalizing on the basis of a single example, there are compelling scientific reasons for worrying that our sample of one may be unrepresentative. And this suggests that cynicism about the prospects for universal biology is premature. An inability to formulate a universal theory of a domain of natural phenomena is hardly surprising if one is dealing with a single sample and that sample is unrepresentative. For in such a case one is in no position to discriminate accidental from essential characteristics, even supposing that the phenomena concerned do comprise a natural kind; and even if they do not, one will be in no position to conclude that they do not. As discussed in the next section, this worry is compounded by the fact that modern biology is based on what we now know is an unrepresentative form of Earth life, namely, complex multicellular eukaryotes.

5.3 Microbes: The Most Representative and Least Well Understood Form of Earth Life

There is overwhelming evidence that life appeared early on our planet. Earth is around 4.5 billion years old. Until about 3.9 bya (billion years ago) it was subject to a "late heavy bombardment" of meteorites (comets and asteroids), some of which were large enough to vaporize the oceans and melt the crust, making it difficult – but not necessarily impossible (Ryder 2002) – for life emerging earlier to survive (Anbar et al. 2001). Scientifically compelling evidence for life appears almost immediately after this period in the form of carbon isotope ratios in 3.8 bya rocks from Isua, Greenland (Mojzsis et al. 1996). These rocks exhibit a skewed ratio of the lighter isotope of carbon (^{12}C) to the heavier isotope (^{13}C) that is very difficult to explain in terms of abiotic processes; metabolic processes exhibit a quantifiable preference for the lighter isotope of carbon over the heavier isotope. Intriguingly, a recent study of carbon isotope ratios in rocks from Jack Hills, Western Australia, suggests that life was present on Earth even earlier, approximately 4.1 bya (Bell et al. 2015).

By 3.5 bya, bacteria-like organisms were well established on Earth. Sedimentary structures closely resembling those produced by living marine stromatolites – many layered microbial mats produced by cyanobacteria found in, for example, Sharks Bay, Australia – date back to around 3.5 bya (Allwood et al. 2009; Noffke et al. 2013). What appear to be fossilized "bacteria" structures are even found in association with them (Wacey et al. 2011, and, more controversially, Schopf et al. 2007). A caveat is in order, however: Even supposing that these microstructures represent the remains of ancient unicellular organisms, it is not clear that they were (as usually alleged) produced by cyanobacteria. No organic material remains and, as will become clear, morphology provides a poor criterion for distinguishing among different kinds of unicellular organisms. The organisms that built these structures may represent an earlier, now extinct but undoubtedly photosynthetic, type of microorganism.

5.3.1 Planet of the Microbes

Until fairly recently, the prokaryote–eukaryote distinction, which is based upon the internal structure of cells, was considered fundamental to biological theory.[3] Prokaryotes are single-celled microbes that lack membrane-enclosed organelles. There are two kinds of prokaryotes: bacteria and archaea. The cells of eukaryotes, which include both multicellular organisms (animals, plants, some protists, and some fungi) and unicellular organisms (some protists and some fungi), have membrane-bound organelles, the most important of which is a membrane enclosed, chromosome containing, nucleus. The origin of the eukaryotic cell plan (and hence the first single-celled eukaryote) is controversial. The most widely accepted account is that eukaryotes are the result of endosymbiotic unions of prokaryotic cells (Koonin 2010), which would make them younger than prokaryotes.[4] There is little evidence of true multicellular animals and plants until around 640 mya (million years ago) in the form of fossils of the mysterious Ediacaran flora and fauna (Shen et al. 2008).[5] In other words, unicellular organisms (most likely,

---

[3] The distinction between prokaryotes and eukaryotes is no longer thought to be phylogenetically (evolutionarily) significant; indeed, Norm Pace recommends that the term 'prokaryote' be struck from the biological lexicon! Nevertheless, so long as one does not take it as theoretically basic, the term is useful for distinguishing the cell structure of archaea and bacteria from that of eukarya.

[4] Not everyone agrees, however; see Kurland et al. (2006) and Woese (1998).

[5] Macroscopic structures dating back 2.1 mya that seem to be from a very primitive form of multicellular life (some type of colonial organism) have been discovered in Gabon (El Albani et al. 2010).

prokaryotes) appeared very early in Earth's history and dominated life for around 3 billion years before "true" multicellular eukarya (with differentiated germinal and somatic cells) emerged. Recent studies indicate that the evolution of complex multicellular eukarya, with interconnected systems of tissues and cell types, requires substantial increases in genetic complexity (to accommodate crucial changes in cell chemistry and physiology required for true multicellularity) under special but as yet poorly understood environmental conditions (Knoll 2011; Rokas 2008). This suggests that, despite their accessibility to human observation, complex multicellular eukaryotes represent a rare and exotic form of Earth life.

Archaea and bacteria are not only much older than eukarya but also environmentally tougher and far more diverse metabolically. They are found in extremely hot environments (inside geysers and black smokers at temperatures up to 121 °C) and very cold environments (polar seas), in highly acidic environments (pH 0.06) and very alkaline environments, as well as highly saline environments (salt lakes), in the stratosphere (15–20 km above the Earth's surface), and at least 3 km below the Earth's surface – environments in which no eukaryote, unicellular or multicellular, is found.[6] For detailed reviews of this literature, see Schulze-Makuch and Irwin (2008, Ch. 4.3), Baross et al. (2007, Sec. 3.2), and Nealson and Conrad (1999). Bacteria have even been found growing on highly radioactive nuclear waste (Chicote et al. 2005).

The metabolic diversity of archaea and bacteria is much greater than that of (unicellular and multicellular) eukaryotes (Madigan et al. 2006). In addition to exploiting the same basic energy-producing metabolic pathways used by eukaryotes (aerobic respiration, oxygenic photosynthesis, alcohol fermentation, lactic acid fermentation), bacteria and (especially) archaea generate energy using a diversity of other biochemical mechanisms. Some metabolize organic substances under anaerobic conditions using a wide variety of oxidants other than oxygen. Others convert solar energy into chemical energy but, unlike plants, do not release oxygen. Some prokaryotes are also able to generate energy from inorganic chemical compounds, including ferrous iron, elemental sulfur, hydrogen sulfide, and even hydrogen gas, that are not used by any eukaryote to produce energy. The diversity of metabolic strategies used by archaea and bacteria helps to explain their ability to survive under a much wider range of environmental conditions than eukarya.

Bacteria and archaea still dominate life on Earth today. They far outnumber multicellular eukaryotes. Locey and Lennon (2016) estimate that the Earth is inhabited by 10^{11}–10^{12} different species of microbes, and the number of individual microbes is far greater.[7] A mere teaspoon of soil can contain up to a billion

---

[6] Tardigrades, an ancient group of microanimals, are able to survive extreme environmental conditions but, unlike many microorganisms, they do not thrive in these conditions; they become dormant.

[7] And viruses in turn outnumber them (Weinbauer 2004). Viruses are found in large numbers in every microbial community and play a central role in microbial evolution as a source of genetic variation.

microbial cells. The number of bacterial cells in the human body is estimated to be at least as great as that of somatic cells (Sender et al. 2016); some researchers believe that it actually exceeds those of human somatic cells by a factor of 10 to one (American Society for Microbiology 2008). The genetic diversity of the archaea and bacteria is also extremely large, but poorly understood (Oren 2004).[8] It is estimated that less than 1% of prokaryotes have been cultivated and described (Hugenholtz et al. 1998). This is not surprising when viewed in the context of the wide range of environmental conditions in which they are found, the great variety of energy resources that they exploit (as "nutrients"), and the finely tuned manner in which they are integrated into complex microbial communities; it is difficult to anticipate, let alone replicate in a Petri dish, the conditions required for them to thrive. Indeed, when in less than a billion years our sun begins to die, slowly expanding into a red giant that will eventually engulf Earth, archaea and bacteria will be the last form of cellular life, hanging on in isolated pools of hot, briny water until the bitter end, approximately a billion years later. In short, archaea and bacteria dominate the history of life on our planet,[9] with delicate, highly specialized, complex multicellular eukarya occupying a slice of less than a billion years in a span of approximately seven billion years.

Viewed in light of these considerations, it is easy to understand why most astrobiologists suspect that unicellular organisms are fairly common in the universe but that complex multicellular organisms are almost certainly very rare (e.g., Ward and Brownlee 2000). The former emerged very early in the history of Earth and can thrive in what are (from a multicellular eukaryotic perspective) extreme environments along several physical and chemical dimensions (temperature, pressure, pH, etc.). They are also able to metabolically exploit a wide variety of energy sources. In contrast, complex multicellular organisms emerged almost three billion years later, and are environmentally fragile, thriving within a narrow range of physical and chemical conditions, and are highly specialized metabolically. Until the Great Oxygenation Event of around 2.4 bya – in which sufficiently large quantities of oxygen, produced as a waste product by photosynthetic bacteria, built up in the Earth's atmosphere and oceans – there was not enough oxygen in the atmosphere and oceans to support the highly efficient, energetically demanding metabolic processes required by true multicellular eukaryotes (Stamati et al. 2011). The upshot is that complex multicellular organisms provide a very poor foundation for extrapolating to all life, wherever and whenever it may be found in the universe. While still subject to the general $N = 1$ worry, the microbial world

---

[8] Due to HGT, the concept of species is highly problematic for archaea and bacteria. I will have more to say about this in Chapter 6.

[9] Along with their acellular viral associates, which, as discussed in Chapter 6, are now understood to be critical to the structure and dynamics of the microbial world (Suttle 2007).

provides a much more promising source of information about the nature of life. Yet, in keeping with the discussion in Section 4.4 about the tendency of scientists to remain committed to inadequate ontologies, biologists persist in trying to accommodate what they are learning about the microbial world into a theoretical framework founded upon complex multicellular eukaryotes.

5.3.2 A Brief History of Misunderstandings and Surprises

As late as the eighteenth century, Carl Linnaeus, the father of modern biological taxonomy, followed Aristotle in bifurcating life into two "Kingdoms," "Animale" and "Vegetabile" (Linnaeus 1758). For a long time, single-celled organisms – which Antonie van Leeuwenhoek "discovered" (described) in the late seventeenth century, using a new microscope of his own devising, and dubbed "animalcules" on the basis of their self-movement – were not viewed as posing a threat to Linnaeus's classification scheme. By the middle of the nineteenth century, however, it was becoming clear that microorganisms are difficult to fit into the classic Aristotelian dichotomy of plant or animal. While moving about like animals, some microorganisms extract energy from sunlight like plants, whereas others switch back and forth, sometimes nourishing themselves like a plant and other times like an animal, depending upon the environment (Sapp 2003, 85). Furthermore, at the time, biological systematics was based primarily on differences in phenotype and morphology (shape, structure, etc.). It is difficult to classify microorganisms in this way, however. Amoebae, which repeatedly and radically change shape under constant environmental conditions, provide especially salient illustrations. Despairing of finding any unity among these bewildering minuscule organisms, Ernst Haeckel proposed consigning microorganisms to a third Kingdom, Protista, which he divided into eight subcategories; one subcategory, Monera, included not only bacteria but also several minuscule eukaryotes. Most biologists, however, rejected his proposal as *ad hoc* and unwieldy, and continued to classify unicellular organisms as either plant (bacteria and fungi) or animal (unicellular eukaryotes) (Sapp 2003, 85–86, 2005).

During the twentieth century, biological taxonomy was driven theoretically by commitment to Darwin's thesis of common descent and empirically by the development of new technologies for exploring phylogenetic relationships. Edouard Chatton is credited with drawing the distinction between prokaryotes (unicellular organisms lacking "true" nuclei) and eukaryotes (organisms, including plants, animals, and some unicellular organisms, whose cells contain a membrane-enclosed, chromosome containing, nucleus); see historian Jan Sapp (2005) for more on the development of this distinction. By the middle of the twentieth century, many biologists were beginning to view the prokaryote–eukaryote distinction as evolutionarily significant. R. H. Whittaker (1969) proposed placing prokaryotes in their own Kingdom "Monera," eukaryotic microorganisms into a separate Kingdom "Protista," and fungi, which (despite their eukaryotic cell plan) differ from plants in ways thought to be evolutionarily significant, into yet another Kingdom "Fungi." At the highest taxonomic level, life on Earth was now divided into five Kingdoms: Animalia, Plantae, Fungi, Protista, and Monera; see Joel Hagen (2012) for more on the historical development of this classificatory scheme. At long last, biological systematics had officially moved beyond Aristotle's bifurcation of life into plant or animal. The history of microbial systematics from the eighteenth century through the first half of the twentieth century provides a good illustration of the staying power (discussed in Section 4.4) of well-entrenched scientific classification schemes in the face of mounting empirical challenges.

The rapid development of powerful new molecular methods for investigating evolutionary relationships among organisms soon challenged Whittaker's classification scheme. It was difficult to determine phylogenetic relationships among the Monera because they all share the prokaryotic cell organization. Emile Zuckerkandl and Linus Pauling (1965) proposed using nucleotide sequences to circumvent this problem. Shortly thereafter, microbiologists made an astonishing discovery. Despite their prokaryotic cell plan, the Monera do not form a natural phylogenetic category. They are divisible into two subclasses – initially dubbed "Archaebacteria" and "Eubacteria" (Woese and Fox 1977) – which are genetically and biochemically quite different. More specifically, archaebacteria differ from eubacteria in their signature ribosomal RNA, genetic machinery, the absence of peptidoglycan in their cell walls, and the occurrence of ether (versus ester) lipids in their cell membranes (Woese et al. 1978). As Jan Sapp (2005, 292) colorfully observes, the difference between archaebacteria and eubacteria is far greater than that between humans and plants.

Accommodating this discovery taxonomically was not easy. Archaebacteria were found to resemble eukaryotes more closely than eubacteria in their genetic (transcription and translation) machinery and to differ from both eubacteria and eukaryotes in their reliance on ether lipids in their cell membranes. In light of these classification challenges, Woese and Fox (1977) recommended a radical restructuring of biological systematics, which was eventually achieved with the addition of a higher taxonomic rank, the Domain, above that of the Kingdom (Woese et al. 1990). All life on Earth was divided into three overarching Domains (Archaea, Bacteria, and Eukarya). Domain Eukarya was subdivided into four Kingdoms: Protista, Animalia, Fungi, and Plantae. The distinction between plant and animal, which since the time of Aristotle had been *the* fundamental distinction in biological classification, was no longer even *a* fundamental distinction (Woese et al. 1990, 4576). The demotion of plants and animals from a central position in biological taxonomy continues to this day. In the context of a growing body of genomic evidence, Williams and colleagues (2013), among others, argue that Domain Eukarya should be abolished and eukaryotes classified as a subgroup of Domain Archaea in a two domain taxonomic system (Archaea and Bacteria).

In sum, what we have learned in recent years suggests that complex multicellular eukaryotes provide an exceedingly poor foundation for theorizing about life. Scientists have discovered that they are an unrepresentative subsample of what may well be an unrepresentative example of life. Viewed in this light, the microbial world provides the most promising source of information currently available to us about life. Yet as Chapter 6 explains, some of the most central theoretical concepts of contemporary biology are still based upon complex multicellular eukaryotes. The challenge for advocates of universal biology is to take seriously the exhortation, posed in Section 4.4, to explore alternative ontologies – to attempt to conceptualize biology from a microbial perspective – for the purpose of formulating more reliable and robust biological generalizations. This will not solve the $N = 1$ problem. But it is surely a better strategy than persisting in trying to base a universal theory of life on an outlier of Earth life that, despite extensive efforts over a very long period of time, has thus far failed to yield scientifically compelling general principles even for familiar Earth life, let alone life considered generally.

5.4 The Problem of Contingencies and the Origin(s) of Life

An important caveat is in order: It is a mistake to think that a universal theory of life will explain everything there is to know about life. As Section 4.4 explained, the job of a universal scientific theory is to reveal the unity behind the diversity in a domain of natural phenomena. But it does not follow from this that such a theory will be able to explain everything there is to know about the phenomena concerned. Actual (individual) phenomena are subject to contingencies falling outside the scope of the theories that subsume them. This is especially true of biology. Darwin's theory of evolution by natural selection, which provides the closest thing we currently have to a universal theory of life, cannot explain, for example, the end-Cretaceous mass extinction, which destroyed an estimated 78% of all species on Earth 66 mya, including, most famously, the nonavian dinosaurs. Most geologists concur that the end-Cretaceous extinctions were triggered by the impact of a gigantic meteorite in a biologically productive, shallow sea off the coast of the Yucatan peninsula. It is widely believed that the impact would not have been as catastrophic had it occurred in the deep ocean.[10] The same is true for less dramatic biological changes, such as

---

[10] Scientists conjecture that the impact of the massive bolide into extensive sulfate deposits present in the shallow sea produced copious quantities of extremely acidic rain and an impact winter lasting for a decade, rendering recovery from the initial impact very difficult for survivors (Powell 1998, 177–179).

the dominance of marsupials over placental mammals in Australia, which is the result of the gradual separation of Australia from Antarctica that was completed around 45 mya; the separation occurred before the marsupials were largely replaced around the rest of the world by placental mammals.

Historical contingencies need not arise from chance events that make it truly impossible to – as Gould (1990, 48) so memorably characterized it – replay the tape of evolution and get the same outcome.[11] Even *supposing* that there were no chance events – that the tape of evolution would yield the same outcome if it were rewound and replayed under the (and this is the important point) very same physical and chemical circumstances – it would be impossible to reconstruct the history of life on Earth on the basis of Darwin's theory alone. For life on Earth has been shaped by an enormous number of poorly understood and unknown physical and chemical conditions lying outside of the scope of Darwinian evolution. These conditions (e.g., the impact of a particular meteorite at a specific time and place) represent contingencies *vis-à-vis Darwin's theory*. As discussed in Section 4.3, not even fundamental physics is immune to the problem of contingencies. The development of our universe (e.g., formation of stars, current distribution of galaxies, and expansion rate), for example, is thought to depend upon unknown contingencies present immediately before the "big bang" of cosmology. These conditions are not included as part of quantum theory or general relativity. Every scientific theory must cope with contingencies falling outside of its scope if it is to be successfully applied to what actually happens in the natural world.[12]

Like the end-Cretaceous mass extinction and other mass extinctions (e.g., end-Permian and current anthropogenic extinctions), the origin of familiar Earth life is a unique occurrence, different from all other origins of life. Chemical and physical characteristics of the early Earth environment undoubtedly played a role in the processes that produced it, and such characteristics are bound to vary from one incipient cradle of life to another. Moreover, because these chemical and physical characteristics are abiotic they lie outside the scope of any future universal biology. The point is no two origins of life will be exactly the same, and the differences between them cannot be explained by appealing only to generalizations about the nature of life.

It is thus puzzling that so many scientists and philosophers believe that there is an intimate, quasi-logical, connection between an account of the origin(s) of life and an account of its nature. Harold Morowitz, for example, counsels researchers

---

[11] Gould's evolutionary contingency thesis is related to but not identical with John Beatty's "evolutionary contingency thesis" (Beatty 1995). As Beatty argues, Gould's primary concern seems to be with contingent details whereas his concern is with contingent generalizations. I will have more to say about Beatty's thesis, which suggests that universal biology is not a viable project, in Chapter 6.

[12] A few physicists (e.g., Tegmark 2014) attempt to circumvent the problem of contingencies for fundamental physics by advocating for a "mathematical universe" that lacks them. There is no empirical support, however, for this proposal.

that "To ask how life originated, we are going to be forced into that intellectual maze of defining 'life'" (Morowitz 1992, 4). In a similar vein, Joyce (1994) and Pace (2001) launch their now classic discussions of the RNA World with brief defenses of the chemical Darwinian definition of life. And Kauffman (e.g., 2000, Ch. 3) interweaves discussions of his version of the SM (Small Molecule) World with arguments for a novel thermodynamic "definition of life." Advocates of autopoietic definitions of life are even more explicit, using their definitions as "blueprints" for constructing models of the origins of life; see, for example, Luisi's (2003, 2006) discussions of "chemical autopoieses" and the "emergence of life." Even Freeman Dyson, who rejects the bifurcation of models of the origin of life into genes-first accounts and metabolism-first accounts in favor of a "double origins hypothesis," appeals to suppositions about the nature of life in defending his conjecture about the origins of life:

> ... life is not one thing but two, metabolism and replication [genetic-based reproduction] ... There are accordingly two logical possibilities for life's origins. Either life began only once, with the functions of replication and metabolism already present in rudimentary form, and linked together from the beginning, or life began twice, with two separate kinds of creatures, one kind capable of metabolism without exact replication and the other kind capable of replication without metabolism. If life began twice, the first beginning must have been with molecules resembling proteins, and the second beginning with molecules resembling nucleic acids.
>
> (Dyson 1999, 9)

While most researchers contend that an account of the origin of life presupposes an understanding of the nature of life, a few reverse the order of priority. Michael Ruse (2008, 101), for example, argues that a satisfactory "definition" of life presupposes an understanding of the origin of life. Along the same lines, a website advertisement for a workshop on the origin of life counsels prospective attendees: "Life at its origin should be particularly amenable to discovery of scientific laws governing biology..."[13] On either approach, however, the role of contingency in the processes giving rise to life is being minimized if not ignored.

The widespread assumption that one can downplay contingency and extract (what amounts to) a recipe for life from an account of the nature of life is almost certainly false. First, as Section 5.4.1 details, the most influential scientific models of the origin(s) of life, the SM World and the RNA World, are beleaguered with worries about contingencies. Furthermore, as Section 5.4.2 argues, it is not in general true that one can infer how to make a material thing from an account of its

---

[13] Carnegie Institution for Science (2015), *Re-conceptualizing the Origins of Life, Workshop*, November 9–13, 2015 (https://carnegiescience.edu/events/lectures/re-conceptualizing-origin-life).

nature, and there is no reason to suppose that life is somehow special in this regard. The upshot is that it is a mistake to base an account of the origin of life very closely on an account of its nature, or vice versa.

5.4.1 A Plague of Contingencies (on Both the SM World and the RNA World)

As Chapter 1 discusses, the SM World and the RNA World are closely patterned on privileged "definitions" of life, respectively, the chemical metabolic definition and the chemical Darwinian definition. Both are plagued with numerous and diverse worries about contingencies – physical and chemical conditions lying outside the scope of the sweeping generalizations about life upon which they are based.[14] According to the SM World, the transition from nonliving chemicals to a primitive living thing coincides with the development of a rudimentary metabolic system. For this reason it is known as a "metabolism-first" model of the origin of life. The focus of the SM World is on explaining how a collectively autocatalytic network of chemical reactions involving small molecules could develop the increasingly complex levels of chemical organization required for sustaining itself by exploiting energy available in the environment (e.g., Kauffman 2000; Segré et al. 2001; Shapiro 2006).

There are a number of widely discussed problems with the SM World. The identity of the small molecules involved is left fairly open. Most versions focus on the building blocks of proteins (amino acids, peptides, and cofactors). Nevertheless, a few researchers, for example, Robert Shapiro (2006), are open to the possibility of metabolisms based upon small molecules of other types, so long as they are capable of generating proto-metabolic processes. On either scenario, however, the chemical reaction networks concerned require mechanisms for coping with a variety of small environmental molecules, not all of which are advantageous and some of which have the potential to inhibit or disrupt their development. To help solve this problem, most versions of the SM World – and as we shall see, the RNA World too – appeal to highly selective natural "containers" of various sorts: Candidates include lipidic vesicles (e.g., Deamer and Dworkin 2005; Luisi et al. 1999; Morowitz et al. 1988), aerosols (Donaldson et al. 2004), and mineral surfaces and cavities (Cairns-Smith 1982; Martin and Russell 2003; Wächterhäuser 1992).

Most defenders of the SM World (and RNA World) treat mechanisms of containment as lying outside the scope of the SM World, which is not surprising since the target of the model is generating a proto-metabolic (in the case of the

---

[14] For an excellent account of the history of scientific thought about the origin(s) of life, see Iris Fry (2000), and for more detail on conceptual problems with contemporary models of the origin(s) of life, see Cleland (2013).

RNA World, genetic) system. A few researchers (Morowitz 1992; Tessera 2011), however, defend a "Lipid World" model of the origin of life, contending that the "primordial ancestor" of life on Earth was an "autotrophic lipidic vesicle." As will become apparent (Section 5.4.2) these conflicting views on the status of mechanisms of containment vis-à-vis models of the origin of life have something in common: They are byproducts of situating models of the origin of life within the scope of accounts of the nature of life.

In addition, the SM World faces the problem of jumpstarting the kind of dynamic organization required for proto-metabolic cycles. As the frequent invocation of "emergence" by proponents underscores, the physicochemical processes that achieve this are not well understood. Indeed, Stuart Kauffman (2000, Ch. 3) conjectures that a new law of thermodynamics for open systems that are far from equilibrium is required to explain the development of metabolism. Unfortunately, he does not provide a precisely formulated statement of this mysterious law, which amounts to a concession that we currently have no idea how a collection of small molecules could self-organize into proto-metabolic cycles capable of further complexification.[15] Most importantly for purposes of this section, the difficulties that worry him could be a result of unknown contingencies present on the early Earth at the time of the origin of life. Kauffman's model of the origins of life is so tightly coupled to his thermodynamic account of the nature of life that he interprets problems with the former as challenges to the latter, and (in keeping with what was said in Section 4.4) opts for fiddling with the laws of thermodynamics. As will become apparent, this is a common strategy. It is a symptom of mistakenly demanding that a theory of the nature of life subsume the origins of life.

In contrast to the SM World, the RNA World holds that life originated with the development of primitive genetic systems, comprising small RNA molecules, capable of undergoing Darwinian evolution. For this reason, it is classified as a "genes-first" model of the origin of life. The RNA World is predicated on the "spontaneous assembly" of a small RNA oligomer capable of catalyzing its own reproduction or, alternatively, the formation of a small pool of diverse, mutually catalytic, RNA oligomers. The focus is on "evolving" longer and more complex RNA polymers with increasingly efficient catalytic capacities, providing a molecular foundation for a genetic system capable of undergoing Darwinian evolution.[16]

The RNA World faces problems analogous to those confronted by the SM World. The expression "spontaneous assembly" marks just as serious a lacuna in

---

[15] Smith and Morowitz (2015) flesh out an intriguing model of the emergence of life in the spirit of Kauffman's suggestions. Like Kauffman's, however, it presupposes a new understanding of the laws of chemical thermodynamics.

[16] Because it currently dominates scientific discussions of the origins of life, the literature on the RNA World is far more extensive than that on the SM World; see Joyce and Orgel (1999) for a scientifically detailed discussion of the RNA World, and Yarus (2011) for a less technical, more popular discussion.

the RNA World as does "emergence" in the SM World. The assembly of an RNA oligomer from precursor monomers (nucleotides) and the synthesis of the latter from basic molecular building blocks (phosphate, ribose, and nitrogenous bases) under plausible natural conditions on the early Earth face major chemical and physical challenges. Moreover, even supposing that these difficulties could be overcome, the resultant small RNA molecules are extremely fragile. As with the SM World, natural mechanisms of containment, ranging from porous mineral surfaces (e.g., Ferris 2006; Martin and Russell 2003) to lipidic vesicles (e.g., Szostak et al. 2001), are invoked to facilitate the required prebiotic chemical reactions and protect their reaction products from degrading processes and side reactions.

The physicochemical difficulties involved in making sense of the abiotic synthesis of RNA oligomers under natural conditions are so serious – as Christian de Duve (1995) and other defenders of the SM World (e.g., Dyson 1999; Shapiro 2006) argue, more serious than making sense of the abiotic synthesis of peptides under natural conditions – that some advocates of the RNA World (e.g., Crick 1981) contend that it was an extremely improbable (one-off) event. Extremely improbable occurrences cannot be replicated in a laboratory setting. The most that science can do is sanction the claim that they are not physically impossible. Moreover, it looks as if more than one improbable event may be required. A single RNA oligomer having the ability to replicate *itself* must somehow give rise to a chemical network of catalytically diverse RNA molecules capable of replicating each other, a tall order indeed.

Because the prebiotic synthesis of an RNA polymer under natural conditions seems so improbable, some advocates of the RNA World postulate a pre-RNA World involving an analog of nucleic acids that was later replaced by RNA. None of the candidates (e.g., PNA, TNA, or GNA) thus far suggested, however, is much of an improvement over RNA (Anastasi et al. 2007). If it is to provide a scientifically compelling account of the origin of life, the RNA World needs to replace "spontaneous assembly" with a combination of chemical reaction sequences under which the assembly of RNA oligomers from some combination of basic molecular building blocks, under credible conditions on the early Earth, turns out to be plausible.

In this light, a few advocates of the RNA World speculate that the first nucleic acids or their monomeric nucleotides may have been synthesized extraterrestrially, perhaps on dust grains surrounding the protoplanetary disk that eventually gave rise to Earth, and in interstellar dust clouds (Ehrenfreund and Cami 2010). This modest form of panspermia contrasts with the more traditional form, which holds that life, in the form of fully functional cells, originated elsewhere and was brought to Earth in meteorites. The latter version of panspermia is implausible insofar as biological cells have not been found in meteorites. The weaker version, on the other hand, is more promising insofar as nucleobases (e.g., uracil) have been found (along with amino acids) in meteorites (e.g., Martin et al. 2008). Thus far, however, neither nucleotides nor nucleic acid oligomers, which provide the most serious bottlenecks in the synthesis of biologically functional nucleic acids, have been found.

What is most striking about these strategies for shoring up the RNA World, in the face of very substantial chemical and physical obstacles, is how tightly wedded they are to a chemical Darwinian account of life. The focus of the RNA World is on the abiotic synthesis of a genetic molecule (whether RNA or an RNA-analog, such as PNA) capable of jumpstarting Darwinian evolution. Difficulties in making good physicochemical sense of how such a molecule could be synthesized on early Earth are taken as grounds for conjecturing that it must have been produced by a highly improbable occurrence or, alternatively, poorly understood extraterrestrial processes. The possibility that a proto-metabolic system might have arisen first and played a supporting catalytic role in the development of a proto-genetic system based on RNA is typically ignored by its advocates. There is a reason for this. Because the RNA World and the SM World are closely based on competing accounts of the nature of life, they are treated as rivals. Yet as discussed in the following section, the generation of a proto-metabolic system before a genetic system is fully compatible with a chemical Darwinian account of the nature of life.

Treating the RNA World and the SM World as rivals has a significant downside. It makes it difficult to explain how the complex cooperative arrangement between nucleic acids and proteins – characteristic of familiar Earth life and critical to the genotype–phenotype distinction of classical Darwinian evolution – arose. The integration of a metabolic-structural system and a genetic-based reproductive system into a unified living complex falls into the cracks between the models. Some advocates of the SM World try to deal with this difficulty by conjecturing that the first hereditary system may have been compositional, as opposed to list-like (e.g., Segré et al. 2000; Shapiro 2006); the basic idea is that a collection of items (in this case molecules) holds the same information as a list-like structure (sequence of nucleobases) encoding them. But this side-steps the central problem, which is how a nucleic acid polymer able to encode a protein-based metabolic-structural system ever got off the ground. Similarly, postulating a community of RNA polymers developing increasingly complex self or mutually catalytic capacities does not reveal how RNA molecules acquired the ability to encode and coordinate the construction and maintenance of a protein-based, metabolic-structural system. As the old saw goes, "the devil is in the details."

5.4.2 The Origin Versus Nature Problem

The tendency of scientists (and some philosophers) to model the origin of life closely on favored accounts of the nature of life reflects a tacit assumption to the effect that one can figure out how something was made by carefully studying its nature. This assumption is false. Most scientific theories have little to say about how the natural kinds that they subsume are made in the messy, uncontrolled world of nature. Consider the mineral quartz. One cannot infer how quartz forms under natural conditions from either its unique macro-mineralogical properties (hardness, chemical inertness vis-à-vis most substances, heat resistance, crystal habit, etc.) or its molecular composition (SiO_2). These properties, especially the latter, supply scientifically compelling answers to queries about the nature of quartz and they are commonly used to discriminate quartz from other minerals, but they do not reveal the physicochemical processes that actually produce it. Quartz is produced naturally on Earth in several ways. It forms in silica rich magma (molten rock) as it solidifies, and also in certain types of hydrothermal environments. Quartz is also produced artificially in industrial autoclaves. Could quartz be produced in still other ways, under physical and chemical conditions very different from those found on Earth? It seems likely.

Quartz is not an exception. Most material substances have multiple physico-chemical origins. Carbon dioxide provides another illustration. In modern chemical theory, carbon dioxide is distinguished from other chemical substances as a unique molecular compound (CO_2). It can be produced abiotically, biologically, and anthropogenically. Natural sources of CO_2 include volcanic outgassing, wild fires, and respiration. It is also produced in laboratory demonstrations and as a byproduct in certain industrial processes. The reactants and chemical pathways involved in some of these modes of production are quite different. In respiration, for instance, carbohydrates are combined with oxygen through a series of chemical intermediaries to produce CO_2 at body temperatures. Some hot springs, in contrast, produce it at high temperatures through the action of acidified water on dolomite (calcium magnesium carbonate). These natural mechanisms of production cannot be inferred from knowledge of the basic chemistry of carbon dioxide. Admittedly, carbon isotope ratios are used by environmental chemists to help identify the sources of various fractions of CO_2 in the atmosphere, for example, biogenic versus volcanic outgassing. But this information provides little insight into the mechanisms producing a particular fraction of atmospheric CO_2.

The considerations just adduced are pertinent for the acrimonious rivalry between supporters of the RNA World and the SM World. As Section 5.4.1 explained, the physicochemical processes involved in the transition to life from inanimate matter depend upon numerous and diverse characteristics of the environment (temperature, pressure, availability of chemical species, possibilities for physical containment, etc.) lying outside of the scope of an account of the nature of life. The upshot is that a metabolism-first theory could be true for the origin of life even supposing that genetic-based reproduction is more fundamental to the nature of life than metabolism. Alternatively, a genes-first theory could be true for the origin of life even though metabolism is more fundamental to the nature of life than genetic-based reproduction. Moreover, under chemical and physical conditions very different from those of early Earth (e.g., Saturn's moon Titan) there might be a variety of alternative causal pathways for achieving metabolism and genetic-based reproduction, some producing the building blocks for metabolism first, others producing the building blocks for genes first, and some involving the codevelopment of building blocks for both. This underscores the danger of modeling the origin of life too closely on an account of the nature of life.

Some researchers seem vaguely aware that a model of the origin of life could cross-cut a theory of the nature of life in the manner suggested. In a famous paper, Leslie Orgel (1998) defends the RNA World on the grounds that a genetic polymer is required to achieve the level of organization necessary for a chemical reaction system to qualify as proto-metabolic. In doing so, he tacitly privileges metabolism as more basic to the nature of life than a genetic system even though metabolism is not being endorsed as the critical step in the transition to life. Analogously, someone who defends the SM World on the grounds that the synthesis of RNA oligomers and their precursors presuppose proto-metabolic networks involving small molecules is implicitly committed to a genetic system being essential to life. The latter is especially true of defenders of the SM World (or, for that matter, the Lipid World[17]) who feel compelled to include a compositional genome in the primordial metabolic ancestor of life on Earth. If metabolism were truly sufficient for life, there would be no need for a genetic system at this stage.

There are still other ways in which models of the origin of life and theories of the nature of life could cut across each other. Someone could consistently believe that metabolism and a genetic system are both necessary for life but that the chemical pathways required for producing one or the other represent the central bottleneck in the transition from chemistry to biology. Kauffman (1995), who explicitly endorses a metabolism-first approach, nonetheless suggests something along these lines when he associates the transition to life from nonliving ensembles of molecules with the development of "complex, non-equilibrium chemical reaction systems ... capable of collective self-reproduction, evolution, and exquisitely ordered dynamical behavior" (p. 84). Alternatively, someone might defend a

---

[17] The term "Lipid World" is used ambiguously, sometimes as designating a rival model of the origin of life and sometimes merely as a prebiotic containment theory (compatible with either the SM World or the RNA World); tellingly, it is not always clear which version is being defended!

hybrid theory of the origin of life in which the physicochemical underpinnings of metabolism and genetic-based reproduction develop concurrently while still being committed to one or the other (but not both) being essential to life.

Lack of clarity over one's views about the nature of life can lead to theoretical ambiguities and confusions over what needs to be included in a theory of the origin of life. Copley and colleagues (2007) provide a good illustration. Although defending a hybrid model of the origin of life, where the complex cooperative arrangement between nucleic acids and proteins begins very early in a chemical reaction system consisting of amino acids and ribonucleotides, they nonetheless classify their account as an RNA World theory. This suggests that they are committed to a genetic-based theory of the nature of life, which they seem to believe also commits them to a genes-first model of the origin of life, despite the fact that they have clearly illustrated that it does not by developing a hybrid model. Vasas et al. (2010) provide an even better illustration when they explicitly critique metabolism-first models on the grounds that "... a basic property of life is its capacity to experience Darwinian evolution" (p. 1470). The point is that even supposing that the capacity to undergo Darwinian evolution is fundamental to the nature of life it does not follow that metabolism-first models of the origin of life are *ipso facto* untenable; metabolism-first accounts may indeed have serious problems of the sort that Orgel (2000, 2004) discusses, but this is a separate issue.

To wrap up, it is a mistake to base an account of the origin of life closely upon a theory of the nature of life, or vice versa. Models of the origin of life and theories of the nature of life may cross-cut each other in different ways. There may be a variety of physicochemical pathways for making life, just as there are a variety of physicochemical pathways for producing quartz and carbon dioxide. If this is so, there will not be a single, context free, explanation for the origins of life, even supposing that life is a natural kind. Instead, there will be alternative explanations, each keyed to pertinent features (contingencies) of the particular environment in which life arises. Viewed in this light, the antagonism between supporters of the RNA World and supporters of the SM World is counterproductive. It engenders a false dichotomy with regard to the physicochemical possibilities for synthesizing life under natural conditions, and hence may hinder the development of scientifically more plausible models of the origin of life on Earth (and elsewhere in the universe as well). Failure to disentangle contingency ridden models of the origin of life from highly general accounts of the nature of life may also produce the illusion that there is no natural unity to the phenomena of life. But even supposing that there are a variety of different physicochemical pathways for making life, it does not follow that life *per se* is not a natural kind. A pluralist stance towards the *origin* of life is fully compatible with a monist stance towards the *nature* of life.

5.5 Concluding Thoughts

Our scientific understanding of life is based upon observations of known Earth life, for they are the only examples of life of which we can be certain. Generalizations about the nature of life, including speculations about artificial life and weird forms of extraterrestrial life, are based upon them. Conjectures about the origin(s) of life are inferred from what we have learned about the chemical composition and molecular architecture of familiar life. Yet as Section 5.2 explained, biochemists and molecular biologists have established that life could differ from familiar Earth life in some modest ways at the molecular and biochemical levels. Moreover, astrobiologists have explored how the basic functions of familiar life (metabolism and genetic-based reproduction) might be realized by molecular compounds based on elements other than carbon under chemical and physical conditions differing from those thought to have been present on the early Earth. The truth is we really do not know how different life could be at the molecular and biochemical levels from familiar life.

As Chapter 1 explained, there is a long history, going back to Aristotle, of taking complex multicellular organisms as the prototypes for life. As discussed in Section 5.3, however, microbiologists have discovered that these organisms are biological outliers – highly specialized, fragile, latecomers to our planet. The most representative form of life on Earth, and almost certainly elsewhere in the universe, is microbial. Bacteria and archaea are far older, environmentally tougher, and metabolically more diverse than multicellular eukaryotes. Earth's microbial world thus provides a much more promising source of information about the nature of life than the world of complex multicellular eukaryotes. There are of course no guarantees that life elsewhere in the universe will closely resemble known Earth microbes. But the likelihood that it will is surely greater than the likelihood that it will resemble complex multicellular eukaryotes. Besides, even supposing that life elsewhere is significantly different from familiar life, a theoretical framework based on Earth microbes will at least provide us with a better understanding of life on Earth.

As Section 5.4 discussed, it is a mistake to demand of a universal theory of life that it be able to tell us everything there is to know about life, including how it originated under natural conditions on the early Earth. On the other hand, the ability to distinguish highly contingent characteristics of familiar life from those that are more basic to life is crucial to being able to develop a universal theory of life. Biologists use the expression "frozen accident" for characteristics (e.g., the genetic code) thought to have arisen from highly contingent conditions present during the origin and early evolution of familiar life. But it applies to many other characteristics of Earth life as well. As an illustration, humans have five fingers and five toes, but there are no reasons to suppose that they could not have had a different number. Birds, for instance, have three digits on their fore limbs (wings) and four digits on their hind limbs. So why do humans have five digits on all four limbs? The standard evolutionary answer is that it is contingent: Mammals evolved from a tetrapod lineage that just happened to have five digits on each limb, and this trait has been passed down unchanged to modern mammals. The point is many characteristics of familiar life seem contingent and, given that we are not able to compare familiar life with other examples of life, it is difficult to say which ones provide the most promising candidates for theorizing about all life.

Basing a general theory of life on Earth microbes will not solve this problem. But it is likely to provide us with a more unifying account of familiar Earth life than one based on multicellular eukaryotes, and depending upon how representative Earth life is (something that we are currently in a very poor position to evaluate), it might even point the way towards a better understanding of life in general. For as Chapter 4 discussed, in the context of illustrations from the history of science, theory construction is typically a slow process. It begins with the development of a set of core theoretical concepts (an ontology) for formulating and, most importantly, exploring generalizations about a domain of phenomena. Whether or not a privileged set of theoretical concepts can support scientifically fruitful generalizations – generalizations that maximize scope while minimizing exceptions – depends upon hitting the right (what I dubbed "Goldilocks") level of abstraction (Section 4.3). To the extent that the microbial world is likely to be more representative of life than the world of rare and exotic multicellular eukaryotes, the former is a more promising place to seek an ontology for a universal biology. As discussed in Section 5.4, however, there is the ever present danger of becoming prematurely committed to what seems to be an intuitively promising ontology for biological theory; the extent to which a set of concepts for life seems intuitively promising will to a great extent depend upon how closely it conforms to our current (eukaryote-centric) beliefs about life, and this is just what we are trying to avoid. The best strategy is thus to explore the capacities of alternative ontologies, formulated in light of our growing knowledge of the still mysterious microbial world, for supporting generalizations that maximize scope (apply to all known Earth life, microbial and, for lack of a better word, macrobial) while minimizing exceptions. In this light, the next chapter (Chapter 6) explores ways in which our eukaryote-centric concepts break down when extended to the microbial world, setting the stage for reconceptualizing biology from a more microbial perspective. The bottom line, however, is that what the program of universal biology really needs is examples of life descended from an alternative biogenesis. The last three chapters of this book explore and evaluate three different strategies for acquiring them: (1) artificially creating novel life right here on Earth (Chapter 7), (2) searching for extraterrestrial life in the absence of a definition of life (Chapter 8), and (3) searching for a "shadow biosphere" (microbes descended from an alternative origin of Earth life, Chapter 9).

6

Rethinking the Traditional Paradigm for Life: Lessons from the World of Microbes

6.1 Overview

As discussed in Chapter 1, Aristotle divided all life into two taxonomic categories, plant and animal, a view that, as Section 5.3.2 recounts, dominated biology until less than two hundred years ago. When one considers that Aristotle's observations were limited to what could be seen by means of unaided human vision, namely, plants, animals, and certain fungi, for example, mushrooms (which he classified as plants), this is hardly surprising. In the seventeenth century, Antonie van Leeuwenhoek, who first observed and described them under a microscope of his own devising, classified microorganisms as tiny animals ("animalcules"). It was not until the mid-nineteenth century that unicellular organisms were placed in their own (a third) taxonomic category, Protista, by Ernst Haeckel. What is surprising is how long Aristotle's classification system survived in the face of mounting empirical evidence that unicellular organisms defy classification as plant or animal.

By the late twentieth century, however, it had become clear that bacteria and archaea are the oldest, most diverse (both genetically and metabolically), and abundant form of life on Earth, and that plants and animals are exotic latecomers to our planet. Yet as this chapter discusses, when it comes to theorizing about life, the latter are not treated as outliers. Following Aristotle, they still serve as the paradigms of life. This suggests that the ostensible disunity of the phenomena of biology, often cited by pluralists against the prospects for universal biology, may be little more than an artifact of a defective (multicellular) eukaryote-centric theoretical framework for reasoning about life. For as Chapter 4 discussed, in the context of examples from the history of science, the manifestation of disunity is more often than not a sign of a flawed theoretical ontology for generalizing about a domain of phenomena, as opposed to evidence that the phenomena are intrinsically disunified.

It is beyond the scope of this chapter to develop a proposal for theorizing biology from a microbial perspective. My goal is the modest one of (i) exploring how some widely accepted concepts and principles, founded upon observations of complex multicellular eukaryotes, may be hindering the development of a more comprehensive understanding of Earth life and (ii) signposting, primarily through the medium of suggestive questions, some intriguing directions for pursuing a more microbe centered theoretical framework for biology. The hope is that this will motivate others to explore the microbial world for more fruitful theoretical concepts for generalizing about life on Earth, and perhaps elsewhere in the universe as well. Whether such efforts will be successful is an open question. What is certain is that such a theory will not be developed if researchers do not pursue it.

6.2 Evolution Viewed Through the Lens of the Microbial World

Many biologists – most famously Richard Dawkins (1976, 1983) – believe that Darwin's theory of evolution by natural selection comes close to providing us with a universal theory of life. This conviction underlies the dominance of Darwinian definitions of life, such as NASA's "chemical Darwinian Definition" (Section 2.2.3), and Darwinian based models of the origin of life, such as the RNA World (Section 5.4). What we are learning about the microbial world, however, is casting doubt upon some widely accepted concepts and principles associated with classical Darwinism.[1] This section focuses on challenges posed by horizontal gene transfer (HGT) – sometimes also called lateral gene transfer (LGT) – to biological thought about the evolution of life on Earth.

In classical Darwinism, heritable characteristics are transferred vertically, from parent to offspring, by means of reproduction. All living things on Earth engage in this mode of gene transfer, which is sometimes called vertical gene transfer (VGT). Another mode of gene transfer, in which organisms transfer genetic material directly (horizontally or laterally) from one organism to another independently of reproduction, was discovered in the twentieth century. HGT is common among bacteria and archaea, and by the end of the twentieth century was recognized as playing a significant role in their evolution (Sapp 2003, Ch. 8). The primary

---

[1] I hasten to add that *nothing* that we are learning about the microbial world is challenging a naturalistic explanation of life, let alone providing support for creationist accounts of the origin of life or the origins of species. Creationists engage in an inexcusable logical lapse (and exhibit a profound misunderstanding of the nature of science) when they construe the original version of Darwin's theory as dogma and take empirical challenges to certain aspects of it as evidence for an "intelligent [divine] designer." As discussed in Chapter 4, scientific theories are incomplete and open to revision in light of empirical and conceptual difficulties, and when this fails they are eventually replaced by better theories. The point of this section is merely to point to some difficulties that the microbial world poses to a conception of evolution based closely upon what we now know are outliers of familiar Earth life, namely, complex multicellular eukaryotes (Section 5.3).

mechanisms of HGT are (i) transformation (uptake of extracellular DNA available in the environment), (ii) transduction (transfer of foreign DNA to an organism via viral infection), and (iii) bacterial conjugation (transfer of DNA between bacterial cells during cell-to-cell contact). By means of these mechanisms, a prokaryote acquires genetic material from a cell other than its parent cell. Some of this foreign genetic material may confer adaptive phenotypic traits, such as antibiotic resistance or the ability to metabolize a certain nutrient, on a microorganism,[2] increasing the odds that the recipient will survive long enough to reproduce and transfer the trait to its daughter cells (via VGT).

At the very least, HGT undermines the classical Darwinian canon that the flow of genetic information is exclusively vertical, from parent to offspring. The question is: How significant is HGT in the evolution of life on Earth? To the extent that genes acquired by means of HGT are evolutionarily relevant only if they are passed on to offspring, HGT might be thought to pose little challenge to the theoretical framework bequeathed to us by Darwin. Natural selection operates upon whatever heritable traits an organism happens to possess, regardless of how they were acquired. If the traits are advantageous the organism is more likely to survive and pass them on to offspring; if they are not advantageous, the organism is less likely to survive and reproduce. In order for HGT to pose much of a challenge to evolutionary theory, the following, widely accepted, assumptions would have to be false: (a) HGT is a minor source of the heritable variation fueling evolution by natural selection and (b) genes acquired by means of HGT are random with respect to the adaptiveness of the traits conferred. Discoveries about the microbial world are raising questions about both assumptions. As the subsections below discuss, this has consequences for some widely accepted biological theses about the origin, development, and diversification of life on Earth.

6.2.1 The Concept of a Biological Species and the Tree of Life

HGT is rare among eukaryotes, especially animals, whose chromosomes are not only sequestered within a membrane enclosed nucleus but which also have special reproductive (germ) cells.[3] Viral infection of germ cells is pretty much the only way that it can occur. Just how extensive HGT is among prokaryotes is controversial. HGT is most common among closely related microorganisms but also occurs among distantly related microorganisms, including Archaea and Bacteria (Boto

---

[2] Much of the genetic material acquired by means of HGT consists of fragments of DNA, and hence is not "genomic" in the sense of encoding viable hereditary information. Nevertheless, some of it does, and is critical to the survival of a unicellular microbe under rapidly changing environmental conditions.

[3] HGT appears to be more common in plants but the mechanisms are more diverse and not as well understood (Gao et al. 2014), which is why this section focuses on HGT in animals.

2010). Moreover, there is evidence that HGT may be far more common than widely believed among bacteria from different genera (McDaniel et al. 2010). Viewed from the perspective of the long history of life on Earth, however, even rare cases of HGT among distantly related microorganisms occur frequently enough to be evolutionarily significant.

As an especially salient illustration, all eukaryotic cells possess mitochondria and some eukaryotic cells (those of plants) possess plastids as well. These organelles contain their own sets of genes which closely resemble those of certain aerobic bacteria (mitochondria) and photosynthetic bacteria (plastids). Following Lynn Margulis (Margulis 1981; Sagan 1967), the widely accepted explanation for this puzzling discovery is that these organelles are remnants of what were once free-living bacteria (Koonin 2010). Around 1.8 to 2.2 bya, an archaean cell is thought to have ingested an aerobic bacterium, creating the first eukaryote cell. Subsequently, one of these hybrid cells ingested a photosynthetic bacterium, creating the first plant cell. Incorporating foreign cells into a host cell as organelles required transferring and integrating genes from the chromosomes of the former to the latter in a horizontal (versus vertical) manner (Timmis et al. 2004). This helps to explain why the chromosomes of eukaryotes contain large quantities of what seem to be archaeal and bacterial genes. In essence, eukaryotes are genetic mosaics consisting of a mixture of genes derived from ancestral archaeal and bacterial sources. As discussed in Section 5.3.2, the core genetic machinery used by eukaryotes for transcription and translation is more similar to archaea than bacteria, which helps to explain why the host cell is thought to have been an archaeon. In this light, some biologists (e.g., Williams et al. 2013) are lobbying for a two domain (Bacteria and Archaea) taxonomic system, where the eukarya are classified as a branch of Archaea.

HGT threatens the highly influential concept of species, which lies at the heart of evolutionary theory. It is not an accident that Darwin titled his magnum opus *On the Origin of Species*. The concept of species continues to play a central role in evolutionary theory. The influential model of a tree of life is a tree of species (Figure 6.1). At its base lies a theoretical cell known as LUCA (Last Universal Common Ancestor), from which all known life on Earth allegedly evolved. Most scientists investigating the origin of life conceive of their task as that of reconstructing the geochemical processes that produced LUCA. The pervasiveness of HGT among prokaryotes makes it difficult, however, to partition them into stable species, which undermines the idea of a tree of life rooted in an ancestral cell. An *E. coli* bacterium, for example, may acquire genes from bacteria that are not very closely related to it, rendering it different enough from other *E. coli* bacteria that classifying them together as members of the same species would be like classifying humans and lemurs as members of the same species (Oren 2009).

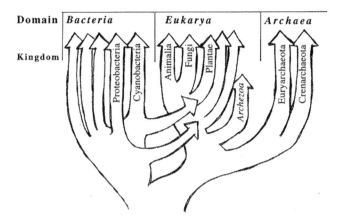

Figure 6.1 The standard model of the tree of life, with a common root from which the three domains of life descend; the diagonal arrows represent the transfer of genes from mitochondria and chloroplasts through endosymbiosis.
(Figure courtesy of W. F. Doolittle, Phylogenetic classification and the universal tree, *Science*, **284**, 1999, Figure 2).

Even more worrisome, from the perspective of classical Darwinism, HGT threatens to sever the organismal (in this case, cell) lineage from the gene lineage. For a bacterium that is the product of extensive HGT will have multiple gene lineages – originating from the many incidents of HGT among its progenitors, not to mention itself – that differ from its cell lineage (which is established through cell division). In view of such considerations, Ernst Mayr (1987), the father of the modern biological species concept, despaired of extending the species concept to prokaryotes: In his words, "only sexually reproducing organisms qualify as species" (p. 145).

Initially it was thought that the "bacterial species problem" (as it is known) could be solved by using small subunits of ribosomal RNA (rRNA) to track bacterial cell lineages (Sapp 2003, Ch. 18). As Section 5.2 discussed, ribosomes, which are the sites of protein synthesis, consist of highly complex, tiny molecular machines found in large numbers in every cell. Some subunits of rRNA are so critical to the ability of a ribosome to translate genetic information into proteins that they evolve very slowly and in a systematic fashion; most mutations render them unable to synthesize proteins, and hence lethal. A good illustration is 16S rRNA, which is present in most microorganisms and is commonly used to differentiate prokaryotic cell lineages from HGT acquired gene lineages. There is evidence, however, that even these "core" ribosomal genes have been subject to HGT and have undergone more mutations than anticipated (Tian et al. 2015), potentially undermining their usefulness for differentiating among cell lineages. Even more problematically, 16S rRNA is too coarse a tool to differentiate among prokaryotes at taxonomic rankings as low as those of species, genus, and family; although the 16S rRNA gene contains subunits of differing degrees of variability, for making finer grained distinctions, there are not enough of them.

Biologists and philosophers of biology respond to the bacterial species problem in two different ways. Some persist in ingenuous efforts to formulate a univocal concept of species; see Ereshefsky (2010) for a review and critique of this literature. A more common strategy, however, especially among philosophers, for example, Dupré (1993), Ereshefsky (2001), Franklin (2007), Kitcher (1984), and O'Malley (2014, Ch. 3), is to defend a pluralist account of species; although they rarely explicitly defend such an approach, many biologists accede to it in practice. Species pluralists contend that there are a variety of equally legitimate ways of classifying organisms into species. As an illustration, consider the biological species concept (BSC) and the phylogenetic (evolutionary) species concept (PSC). The BSC, which is usually attributed to Mayr (1942), identifies a species as a group of organisms capable of interbreeding and producing fertile offspring. The PSC, on the other hand, identifies a species as a group of organisms that share a common ancestor distinct from that of other organisms. Sometimes these categories overlap and sometimes they do not. Donkeys, for example, satisfy both the BSC and the PSC. Mules, on the other hand, do not satisfy the BSC since they are incapable of producing fertile offspring. The BSC and PSC are difficult, however, to extend to the microbial world. Prokaryotes do not (technically speaking) interbreed (sexually reproduce), although bacterial conjugation is sometimes characterized as a form of "sex." And as we have seen, the prevalence of HGT among prokaryotes makes it difficult to group them into species having unique, common organismal (cell) ancestors. These considerations suggest that it will be very difficult, if not impossible, to formulate a truly universal (biological or phylogenetic) concept of species.

From a pluralist perspective the bacterial species problem is not very serious. Biologists are free to formulate special purpose species concepts as needed, and indeed many have done so; see Ereshefsky and Pedroso (2016) for a discussion of the many different concepts of species, some designed for prokaryotes and others for eukaryotes, that have been proposed. Moreover, just as species concepts designed for multicellular eukaryotes sometimes cross classify them into different categories, microbial species concepts can cross classify microorganisms into different species. The question is whether proliferating numerous, sometimes overlapping and sometimes disjoint, special purpose, species concepts is fruitful for biological theory. By sanctioning disunity, species pluralism discourages searching for theoretical unity. On the other hand, as just discussed, efforts to formulate a univocal concept of species have been notoriously unsuccessful. What choice, other than proliferating species concepts, do biologists have?

Section 4.3 suggested another option: Seek new, more unifying and scientific-ally fruitful, theoretical concepts from the microbial world. For as was the case with the now obsolete concepts of phlogiston and the luminiferous aether, the difficulty may lie with the very concept of species: The concept of species may be a relic from a time when VGT was erroneously deemed the dominant mode of gene transfer for understanding the evolution of life on Earth. Growing evidence that HGT has played important roles in the evolution of multicellular eukaryotes underscores this point. The genomes of animals and plants are littered with retro-viral DNA (acquired from viruses via HGT); 5–8% of the human genome, for example, consists of retroviral DNA (Belshaw et al. 2004). There is mounting evidence that some of this foreign DNA – originally viewed as either junk or pathogenic – played a significant role in the diversification of animals (Boto 2014) and plants (Yue et al. 2012). It is now widely acknowledged that the evolution of the mammalian placenta, for instance, depended upon a gene transferred horizon-tally from a virus (Dupressoir et al. 2012).

To wrap up, if HGT plays as central a role in the evolution of life on Earth as many microbiologists are beginning to suspect, then the concept of species (and other taxonomic categories as well) lacks the theoretical import traditionally assigned to it in biological theory. Salvaging the species concept in a pluralist manner – by retaining the term 'species' (come what may) and giving it a multiplicity of special purpose meanings – is unlikely to be of much help in rectifying this situation. In truth, as discussed in Section 4.4, it is likely to frustrate the development of a scientifically more inclusive theoretical framework for reasoning about life. As Section 4.3 counsels, a more promising approach is to seek novel theoretical concepts, for supporting generalizations about life, from among the most representative form of life on Earth, namely archaea and bacteria. Along these lines, some microbiologists recommend that the widely accepted tree of species be replaced by a collection of networks of genomes (Figure 6.2); see, for example, Dagan et al. (2010), Doolittle (1999), and Doolittle and Papke (2006). Insofar as a network lacks a base, this proposal dispenses with the organismal concept of LUCA. Somewhat ironically, however, some proponents of the network model of the evolution of life on Earth advocate retaining LUCA by construing it as a collection of genomes (versus a microorganism), illustrating yet again the unfortunate tendency of scientists to salvage venerable but nonetheless problematic concepts by reinterpreting them.

6.2.2 Is Lamarck Hiding in the Shadows?

This brings us to the second evolutionary doctrine mentioned earlier, namely, that heritable variation, which includes genes acquired via HGT, is random with respect

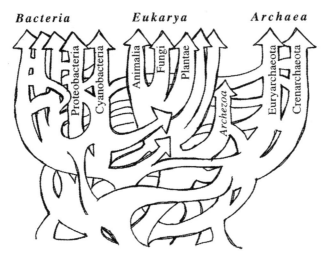

Bacteria **Eukarya** **Archaea**

Figure 6.2 Ford Doolittle's alternative to the tree of life, which takes account of the flow of genes via HGT across all three domains of life. The result looks more like a bush or web than a tree insofar as it lacks a trunk or root.
(Figure courtesy of W. F. Doolittle, Phylogenetic classification and the universal tree, *Science*, **284**, 1999, Figure 3).

to its adaptive significance. Questioning this assumption is widely frowned upon as cozying up to Lamarckian evolution, which is viewed as tantamount to sacrilege by many biologists. Admittedly, what many contemporary biologists and philosophers dub "Lamarckian evolution" is disparaged for good reason. It requires a mechanism for directly encoding adaptive traits acquired by an organism during its lifetime into its genome. Given our current understanding of the molecular basis of heredity, it is difficult to conceive of a natural mechanism for reverse engineering a genome in this manner.

As historian Jessica Riskin (2018) explains, however, the hostility frequently directed towards Lamarck is based on the mistaken notion that his ideas were unscientific and opposed to those of Darwin. Yet Lamarck, like Darwin, was a naturalist and based his theory of evolution on careful observations. Moreover, he preceded Darwin in rejecting Aristotle's view that species are eternal and unchanging in favor of the idea that they evolve. Darwin himself accepted Lamarck's mechanism of the inheritance of acquired characteristics, which he referred to as "use and disuse"; as Riskin discusses, this mechanism is included in every edition of *The Origin of Species* (Darwin 1859). According to Riskin, the person responsible for purging Lamarckian ideas (about the inheritance of acquired characteristics) from Darwinian evolution was August Weismann. Weismann's interpretation of Darwin's theory became the standard version, which to this day holds that the core process driving evolution is random genetic variation acted upon externally by natural selection. In other words, "classical Darwinism" is really Weismann's interpretation of Darwin's theory.

Neither Lamarck, Darwin, nor Weismann understood the physical basis of heritable variation, which they all recognized is required for evolution to occur. It was not until the mid-twentieth century, with the discovery of the structure of DNA, that Weismann's notion of heritable variation was fleshed out physically in terms of random changes to DNA through mutation.[4] In recent years, mechanisms of heritable variation with a distinctively Lamarckian flavor have been discovered, most notably, epigenetic inheritance, which involves environmentally induced, heritable changes in gene expression that do not involve changes to DNA, and, most importantly for the discussion in this section, the direct transfer of DNA from one organism to another, independently of reproduction, by means of HGT; see Jablonka and Lamb (2005) for a discussion of these and other "Lamarckian" mechanisms. The upshot is that biologists have actually discovered that no encoding of acquired characteristics is required for the inheritance of acquired characteristics.

The Lamarckian character of HGT has been acknowledged by some scientists, for example, Goldenfeld and Woese (2011) and Koonin and Wolf (2009). Certain genes found in prokaryotes are much more prone to mutation than others. The rates of mutation of these "mutator genes" are influenced by environmental factors. Under stressful conditions, such as exposure to antibiotics or depletion of a critical nutrient, they skyrocket in a process known as hypermutation. Genomic material acquired from hypermutating mutator genes by means of HGT during an environmental crisis are more likely (than truly random mutations, distributed arbitrarily across genes) to confer selective advantages upon their recipients. Mutator genes are found in plasmids, small DNA molecules that are able to replicate independently of a cell's chromosomal DNA. Plasmids typically carry specialized genes for coping with or exploiting environmental conditions. Replication and repair of DNA in plasmid mutator genes becomes systematically prone to error under adverse environmental conditions, significantly increasing the likelihood of finding an adaptive solution to the stressor. Cells carrying these potentially life-saving mutations transfer them to other cells via HGT either by means of bacterial conjugation (directly transferring them from cell to cell) or by transformation (releasing them into the environment for uptake by other cells). Interestingly, bacteria have been discovered "harpooning" environmental DNA using tiny fibers known as pili (Ellison et al. 2018), raising the question of whether they might be selecting certain genes over others for inclusion in their genomic repertoire.

---

[4] Mutation is the core mode of heritable variation insofar as it is the mechanism by which DNA is directly modified. The other sources of heritable variation, drift and recombination (gene shuffling), act on genetic material that is already available.

Exposure to environmental stress does not increase spontaneous mutation rates in chromosomal DNA, which is much more likely to be fatal because chromosomes carry genes responsible for core biological functions. Instead, stress-induced mutagenesis selectively targets plasmid genes for increased spontaneous mutation rates while (in essence) "protecting" chromosomal genes. The upshot is that the genomic products of stress-induced mutagenesis are not truly random with respect to the adaptiveness of the traits conferred. On the other hand, mutations arising in plasmid DNA are random with respect to their adaptiveness. It is the sheer number of mutations produced spontaneously in genes specialized for dealing with adverse environmental conditions that increases the likelihood that a bacterial colony will be able to cope with a particular stressor.

As a result of environmental stress, bacterial cells die at a more rapid rate, releasing plasmid DNA into the environment, in the process making an enormous number of genomic variations available for "uptake" (via transformation) by other members of the community. Many cells will receive mutated DNA that has no benefit vis-à-vis the stressor. Some bacteria, however, will be lucky. They will acquire a beneficial trait for coping with the stressor. Intriguingly, there are mechanisms for minimizing the accumulation of deleterious mutations acquired by means of HGT (Galhardo et al. 2007). As a result, bacteria acquiring an adaptive trait are much more likely to survive and pass it on to their daughter cells, which, in turn, will survive and pass it on to their daughter cells. In this manner, a beleaguered bacterial colony may adapt much more rapidly to an environmental stressor than it could using the classical (Weismannian) Darwinian mechanism for heritable variation, namely, truly random mutations (distributed arbitrarily throughout an organism's genome) that are transferred vertically, from parent to offspring, during reproduction.

The extent to which stress-induced mutagenesis can be characterized as Lamarckian depends upon what is meant by "Lamarckian evolution." Koonin and Wolf (2009) cautiously refer to it as "quasi-Lamarckian." Some biologists view the inheritance of acquired characteristics as opposed to natural selection, but this is a mistake. Natural selection operates on genes acquired by means of HGT in the same way that it operates on genes acquired vertically through reproduction. If genes acquired via HGT confer a selective advantage on the recipient, the likelihood that it and its offspring will survive is increased; if they do not, the chances of survival are not increased and may even decrease. The point is stress-induced mutagenesis casts doubt primarily upon Weismann's account of the randomness of heritable variation, not on the role of natural selection in evolution. On the other hand, it seems clear that stress-induced mutagenesis is Lamarckian in the sense of providing an authentic case of the inheritance of acquired characteristics: It speeds up the process of adaptive evolution by conferring heritable traits on an organism that are (in effect) prescreened for their adaptive potential. Viewed in this light, Koonin's and Wolf's characterization of the process as quasi-Lamarckian seems overly cautious.

Koonin and Wolf discuss other mechanisms involving HGT, which they argue provide even better examples of the nonrandom acquisition of adaptive heritable characteristics. The most intriguing case is the antiphage CRISPR-Cas defense system in archaea and bacteria, which is about as close to reverse genomic engineering as one can get. A bacterium "learns" how to defeat a phage (virus) attack and transmits this information in its genome to its daughter cells; the biochemical details are beyond the scope of this chapter. The fact that they view this as a more promising case than stress-induced mutagenesis underscores the widespread but mistaken belief that the inheritance of acquired characteristics requires that the characteristics concerned be encoded directly into an organism's genome during its lifetime.

The discussion above raises an interesting issue for our understanding of the history of life on Earth. In multicellular eukaryotes, spontaneous chance mutations must occur in specialized germ (versus somatic) cells in order to be passed on to offspring through reproduction. Under conditions of rapid environmental change, multicellular eukaryotes whose germ cells undergo a beneficial mutation are unlikely to survive long enough to pass it on to their offspring because they themselves will not acquire the trait. In contrast, HGT coupled with stress-induced mutagenesis increases the likelihood that some members of a microbial community will acquire the needed characteristic directly, allowing them to survive *and* produce offspring that inherit it. In short, the mechanisms of evolution exploited by bacteria and archaea enhance their ability to survive rapidly occurring, deleterious environmental changes.

In this context, one cannot help but wonder whether an increasingly stable environment could be in part responsible for the sudden appearance of true multicellular organisms in the fossil record around 640 mya years ago. Was the earlier environment of Earth, which was subject to much more extreme events than those of the last 600 million years,[5] too unstable for complex multicellular life to gain a foothold? One cannot help but also wonder whether mechanisms of

---

[5] The conjectured Late Heavy Bombardment, in which meteors as large as moons crashed into Earth, evaporating the oceans, ended around 3.8 bya; the extent of this event has recently become controversial, however (Mann 2018). The Great Oxygenation Event, which occurred around 2.4 bya, is thought to have wrought havoc on (then dominant) anaerobic microorganisms, for whom oxygen is a poison (Sosa Torres et al. 2015). The Earth is also thought to have experienced a series of "snowball" Earth episodes prior to 650 mya in which the entire planet was encased in ice (Hoffman and Schrag 2002; Kirschvink 1992); whether the surfaces of the oceans were frozen solid or just slushy is controversial but it is widely agreed that they were covered in ice. Nothing as extreme as these events (and that includes the great Permian Extinction of around 252 mya) has occurred since the Cambrian Explosion of around 540 mya, which was immediately preceded by the sudden appearance of true multicellular organisms (the Ediacaran flora and fauna) in the fossil record.

evolution on worlds different from Earth might differ in even stranger ways from those with which we are familiar.

In sum, evidence that HGT has and continues to have a major impact on the evolution of all life on Earth is mounting. This has consequences for some venerable evolutionary concepts and theses, including the highly influential model of the history of life on Earth as a tree of species rooted in an ancestral micro-organism and the widespread view that the acquisition of novel genes is random with respect to the adaptiveness of the traits conferred.[6] Some microbiologists have explored alternatives such as jettisoning the concept of species and reconceptualiz-ing the history of life on Earth in terms of a network of genomes or accommodat-ing quasi-Lamarckian ideas in certain aspects of evolutionary theory. These efforts represent much needed but nevertheless fairly modest revisions to evolutionary theory. There are undoubtedly even bolder possibilities that have yet to be con-sidered, let alone explored. In this context, one is reminded of the state of physics before Newton; see Section 4.2. Scientists struggled to formulate a universal theory of bodies in motion using fairly intuitive concepts of body, such as impetus (Aristotelians), occupied volume (Descartes), and weight (Galileo). It was not until Newton placed a very counterintuitive concept (inertia, or passive resistance to a change in motion), based on observations of the motions of celestial objects, which (unlike familiar terrestrial objects) never cease moving, at the core of a theory of motion, that scientists finally acquired principles of motion that were truly general, encompassing the motions of projectiles, pendulums, falling objects, planets, the tides, etc. Could the disunity plaguing contemporary biology be a consequence of factors analogous to those that afflicted physics before the advent of Newton's theory, namely, a deeply flawed theoretical framework for generalizing about life based upon observations of an unrepresentative sample of Earth life? The next section continues to explore this theme.

6.3 The Living Individual Viewed Through the Lens of the Microbial World

Biology recognizes many different types of entities. Some illustrations are mito-chondria, genes, cells, kidneys, viruses, a bacterium, a rabbit, a spruce tree, an

---

[6] I have not explored another, even greater, challenge to the latter thesis, namely, the deliberate manipulation of genes by human scientists for the purpose of designing organisms adapted to what would otherwise be deleterious environmental conditions (e.g., genomically modified soybeans resistant to weed killers), which is of course even more clearly Lamarckian in character than HGT insofar as it involves mechanisms for directly encoding adaptive traits into the genome of an organism. My discussion is restricted to "natural" mechanisms of evolutionary change because I am concerned with the origin and evolution of life via natural (versus artificial) processes. Still, one could argue that such cases provide even more reason for expanding evolutionary theory to include Lamarckian mechanisms.

aspen grove, a buffalo herd, a honey bee colony, and a species. From a theoretical perspective, however, the most significant is the living individual (basic unit of life), as opposed to components, varieties, or collections of living things. The term 'biological individual' is used in different ways by biologists and philosophers of biology. It is sometimes used synonymously with 'living individual'. Some researchers, however, use the terms 'biological individual' and 'Darwinian individual [unit of selection]' interchangeably but steer clear of identifying Darwinian individuals as living individuals, presumably because not all Darwinian individuals (e.g., genes and viruses) seem to qualify as living things. To avoid confusions such as these, this section restricts the term 'biological individual' to living individuals.

Most contemporary accounts of biological individuality – see Clark (2010), Ereshefsky and Pedroso (2015), and Godfrey-Smith (2009, 2013) – are Darwinian, typically elaborating on Lewontin's (1985) characterization of natural selection in terms of heritable variation and differential fitness. In essence, for these philosophers, living things are units of (natural) selection. The Darwinian view of biological individuality has not always dominated scientific thought about life. Stereotypical organisms, most notably animals, are paradigmatic biological individuals, and as Godfrey-Smith (2013, 25) observes, the traditional concept of organism is metabolic. Animals are both metabolic and Darwinian individuals. As Godfrey-Smith discusses (2013, 28–29), however, not all Darwinian individuals are metabolic individuals and, somewhat surprisingly, not all metabolic individuals are Darwinian individuals. The upshot is that the organismal and Darwinian concepts of biological individuality are not coextensive, raising the question of whether being a Darwinian individual is truly more basic to biological individuality than being a metabolic individual.

The question of whether biological individuality is metabolic or Darwinian glosses over an earlier, *prima facie* more basic, notion of biological individuality. As discussed in Section 1.2, on Aristotle's account of biological individuality, *autonomy* – the capacity of living things to do things (eat, grow, move, reproduce, etc.) *on their own* – distinguishes living from inanimate individuals (such as a rock). Aristotle's notion of autonomy undergirds both metabolic and Darwinian accounts of biological individuality. On a metabolic view, biological individuals sustain and repair themselves as metabolically coherent wholes by extracting and transforming energy available in the environment. On a Darwinian view, biological individuals contain the within themselves machinery for the reproduction of a new individual of the same kind, which is a prerequisite for being an evolutionary unit of selection. Aristotle's concept of biological autonomy plays a role in many other areas of modern biology. As an illustration, the discovery that some cells (e.g., bacteria) are "free living" – able to perform the allegedly basic functions of life by themselves – lies behind the principle, enshrined in "cell theory" that the cell is the minimal unit of life. It also lies behind the controversy, among origin of life researchers, over whether the first living things were metabolic individuals (Small Molecule World) or Darwinian individuals (RNA World); see Section 5.4 for more on these models.

Aristotle analyzed biological autonomy in terms of a special form of causation, self-causation, unique to living things. Although the Aristotelian thesis that life involves a unique form of causation is no longer accepted, most contemporary biologists nonetheless accept the associated idea that biological individuality presupposes a distinction between self and other, and cash it out in terms of enclosure by a physical barrier, providing a definite inside (self) and outside (other).[7] Animals and plants provide the model. What goes on inside the epidermis of an animal or plant – the activities of cells, tissues, and organs – maintains its functional and structural integrity, as well as orchestrating its reproduction. Unicellular microorganisms (bacteria, archaea, and single-celled eukarya) are self-metabolizing and self-reproducing in a very similar manner. Their life conferring, metabolic and reproductive, processes occur within cell membranes and cell walls.[8]

The neo-Aristotelian doctrine that biological autonomy requires confinement of basic biofunctions within physical barriers retains a powerful grip on the minds of biologists. Most theories of the origin of life, for example, treat physical encapsulation as a necessary condition for the emergence of the chemical reaction networks required for bootstrapping whatever biological function(s) – genetic-based replication (RNA World) or metabolism (Small Molecule World) – is theorized as foundational to life (Popa 2004, Ch. 3). What we are learning about the microbial world, however, is challenging this neo-Aristotelian concept of biological autonomy, raising the possibilities of yet unrecognized forms of life.

Viruses provide the most widely discussed candidate for an acellular form of life. Consisting of a protein coated, single or double stranded, DNA or RNA molecule, viruses can neither metabolize nor reproduce "on their own," and for this reason are traditionally excluded from the category of living things (Moreira and López-García 2009; Nasir and Caetano-Anollés 2015; Van Regenmortel 2016). Nonetheless, viruses actively initiate and direct their own reproduction by invading and commandeering the genetic and metabolic machinery of cells available in the external environment. Viruses are also Darwinian individuals in the

---

[7] As discussed in Section 1.2, Aristotle believed that biological "self-causation" is different in kind from nonbiological linear ("efficient") causation. While neo-Aristotelian "vitalist" notions about life have fallen out of favor among contemporary biologists (see Section 1.3), a neo-Aristotelian residue still lurks in the widely accepted principle that the life conferring functions of biological individuals must be internal to them.

[8] All cells have membranes. The membranes of some (but not all) unicellular organisms are enclosed by yet another barrier, known as a cell wall.

sense that they respond to the pressures of natural selection as coherent wholes. In light of these considerations, one cannot help but wonder whether it is a mistake to require of a living thing that it carry out the basic functions of life inside its own skin (so to speak). Some researchers (e.g., Bamford et al. 2002; Forterre 2016) believe that it is a mistake, and contend that viruses should be considered an acellular form of life. Not everyone agrees, however; see Koonin and Staroka-domskyy (2016) for a review of the debate over whether viruses are living or nonliving individuals.

From a philosophical perspective, the allegation that viruses should be counted as living things rests upon their agent-like behavior. Agency can be thought of as a form of autonomy that is initiated internally but operates externally. By harnessing the metabolic and reproductive machinery of cells found in their environment, viruses act externally as agents for themselves. As the next three subsections discuss, viruses are not the only biological entities posing challenges to neo-Aristotelian assumptions about the nature of biological autonomy.

6.3.1 Is the Host–Microbiome Complex (Holobiont) a Living Thing?

As Gilbert and colleagues (2012) and Dupré and O'Malley (Dupré and O'Malley 2007a; O'Malley 2014, Chs. 5 and 6) observe, the supposed metabolic and reproductive autonomy of stereotypical organisms, which (historically speaking) lies at the core of the neo-Aristotelian concept of the living individual, is illusory. Complex multicellular organisms depend upon microbes – especially bacteria but also archaea, viruses, algae, fungi, and sometimes even protists – for many biological functions, from metabolism and reproduction, to development and immunity. The relationships formed between multicellular organisms and their microbial symbionts are intimate and reciprocal; the health and sometimes even life of the former depends upon the activities of the latter, and vice versa.

Legumes (e.g., peas and beans) provide a well-worn illustration. Nitrogen fixation is the process whereby living things convert nitrogen in the Earth's atmosphere into a biologically active form; nitrogen compounds are required for the biosynthesis of some of the basic molecular building blocks (e.g., nucleosides and amino acids) of life. Only bacteria and archaea are capable of "fixing" nitrogen. Legumes acquire biologically accessible nitrogen by (literally) culturing rhizobial bacteria found in the soil around their roots. As sprouts, they collaborate with these bacteria in constructing root nodules, which the bacteria colonize. Root nodules protect and nurture the bacteria by removing free oxygen, which damages the bacterial enzyme involved in nitrogen fixation, and supplying them with organic compounds. Bacteria living in root nodules reciprocate by supplying their hosts with fixed nitrogen.

The intimate cooperative relationship between legumes and rhizobial bacteria is but one of many examples of codependencies between multicellular eukaryotes and microorganisms. Wood is the major source of food for termites but they cannot digest it on their own. Protozoans in their gut digest it for them. Without the aid of their protozoan symbionts a termite would starve. Bacteria in the testes and ovaries of parasitic wasps alter the reproduction of their hosts, for example, allowing infected females to give birth without the participation of males (Schilthuizen and Stouthamer 1997). Studies of mice suggest that bacteria (*Bifidobacterium infantis*) may play a role in reducing anxiety (Foster and Neufeld 2013). Fermentation products of microbiota inhabiting the anal glands of carnivorous mammals play important roles in social interactions such as territory marking and recognition of mate and offspring (Theis et al. 2013). And the list goes on; see Douglas (2018) for a review of the literature.

From an anthropocentric perspective, an even more provocative illustration is the human body, which (like that of all multicellular eukaryotes) is infested with an enormous variety of microbes occupying specialized bodily niches, from the surface of our skin to inside our tissues and organs, most famously our intestines. Microbial symbionts perform many functions important to human health (Bull and Plummer 2014; Douglas 2018, Ch. 3). Microorganisms in the human gut secrete enzymes that are critical for breaking down complex carbohydrates found in fruits and vegetables. Moreover, the gut's immune system and its microbial community co-develop (Planer et al. 2016). Many diseases (e.g., inflammatory bowel disease and psoriasis) and physiological conditions (e.g., obesity) are being linked to an "unhealthy" microbiome (Singh et al. 2017; Wang et al. 2017). Human behavior is also being linked to our microbial inhabitants. Risk taking behavior, which is advantageous in many human activities, such as becoming a successful entrepreneur, has been linked to infection with a protozoan (*Toxoplasma gondi*) present in the feces of domestic cats (Johnson et al. 2018). In short, there is hardly an aspect of human biology that is not affected by our microbial inhabitants, and the human body in turn provides its microbiota with a protected, nutrient rich environment.

Many of the microorganisms comprising the human microbiome coevolved with the human body. Human milk contains components, indigestible to human infants, that are specifically designed to feed gut bacteria (Yamada et al. 2017). Some beneficial bacteria even induce the human body to release chemicals that kill competitors from different lineages (Cash et al. 2006). The mode of transmission of the microbiome from parent to offspring is not, however, through chromosomes or germ cells but instead through inoculation. There is intriguing (but controversial) evidence that colonization begins when the fetus is still in the uterus (Aagaard et al. 2014). As the infant passes through the birth canal microbes are transferred from mother to child (Mueller et al. 2015). After birth the child receives additional microbes through her mother's milk (Hunt et al. 2011) and interactions with things in her environment (Costello et al. 2012). The microbiomes of other multicellular eukaryotes exhibit similar evolutionary and developmental interdependencies, being reconstituted in each generation through a gradual process of inoculation as the host's body develops (Douglas 2018, Ch. 8).

In light of the intimate metabolic, developmental, and evolutionary relationships between a host and its symbiotic microbes, the question arises: Should the host–microbiome complex – commonly (and suggestively) referred to as the "holobiont" – be theorized as a single biological individual, as opposed to an aggregate of separately living, interacting individuals? Answering in the affirmative, Gilbert and colleagues (2012) and Dupré and O'Malley (Dupré and O'Malley 2007a, 2007b, 2009; O'Malley 2014, Ch. 5) propose reconceptualizing host–microbiome complexes as *heterogenomic* multicellular organisms, that is, organisms whose cellular components do not (like the somatic cells of stereotypical multicellular organisms) share the same genome.

On Godfrey-Smith's (2009, Ch. 4) account of biological individuality, however, most holobionts do not qualify as living individuals because they are not reproducers in the requisite, traditional, evolutionary sense.[9] In order for natural selection to operate directly on a biological entity it must reproduce autonomously as a unified whole. Holobionts do not collectively reproduce in this manner. The individual microorganisms comprising a holobiont's microbiome reproduce independently of one another and their host. Moreover, they are transferred to the host's monogenomic, somatic celled, offspring separately, through a gradual, piecemeal, process of inoculation. As a result, despite their remarkable metabolic unity, holobionts cannot be characterized as participating in natural selection as cohesive reproductive units, and hence fail to qualify as Darwinian individuals under the traditional concept of biological reproduction. Insofar as most contemporary biologists and philosophers consider being a unit of selection as *at least* a necessary condition for being a biological individual, this poses a potential hurdle for the claim that holobionts are living individuals.

Seeking a way around this difficulty, some philosophers of biology – for example, Dupré (2012), Lloyd (2018), and Zilber-Rosenberg and Rosenberg (2008) – reject the identification of Darwinian individuals with reproducers. Amending (in different ways) David Hull's (1980) highly influential distinction between interactors and replicators,[10] these philosophers acknowledge that

---

[9] He cites a few exceptions involving single symbiont–host relationships, e.g., the aphid–*Buchnera* (bacterial) symbiosis.

[10] On Hull's (1980) account, natural selection requires *both* interactors and replicators but they need not be the same entity. An interactor is an individual which interacts, as a whole, with its environment in such a way that replication is differential. A replicator is an entity which transfers its structure "largely intact" from one

holobionts are not reproducers *per se* but nevertheless contend that they are units of selection in virtue of being the entities that interact, as wholes, with their environments in such a way that reproduction is differential. On their view, it is the interactive (metabolic) complex (holobiont versus hologenome) which responds to the forces of natural selection as a coherent whole, and hence qualifies as a Darwinian individual. As Elizabeth Lloyd (2018) counsels, in a discussion of holobionts as units of selection, "[i]interactors need not be replicators nor reproducers, and should not be mixed up with them."

It is important to keep in mind that Lloyd and fellow travelers are not denying that Darwinian individuality involves reproduction. Without the transmission of heritable traits from one generation to the next, natural selection (and hence evolution) will not occur. They are rejecting the claim that an interactor must reproduce autonomously, as a unified whole, in order to qualify as a Darwinian individual. It is clear that host–microbiome complexes are not internally integrated in such a way that the host and its microbiome can reproduce together as a combined unit. Their solution is to extend the traditional concept of reproduction to include cases in which many of the parts (microbiota) of a biological individual (holobiont) are reproduced independently and gradually transferred to the next generation via cell migration. What matters is not that the complex reproduces itself as a cohesive unit but that the hologenome (collection of host and microbiota genomes "coding" for its traits) is transmitted with enough fidelity between ancestral and descendant holobionts. As the next two sections discuss, however, there are microbial consortia (biofilms and "rock-powered" ecosystems) exhibiting organism-like metabolic unity which fail to "reproduce" in even this extended fashon.

6.3.2 Biofilms: Aggregates of Cells or Living Individuals?

Biofilms are microbial communities that form on physical surfaces, both nonliving (e.g., rocks in streams and mineral crusts in hydrothermal vents) and living (e.g., mammalian teeth and roots of plants). Some biofilms consist of the same variety ("species") of microorganism. Others are very diverse taxonomically, consisting of bacteria, archaea, viruses, fungi, and microeukaryotes. The boundaries of biofilms are not delineated – in the manner of individual cells and multicellular eukaryotes – by physical barriers. Instead, they are vague and fluid, expanding and contracting in response to changing environment conditions (Stubbendieck et al. 2016).

---

generation to another (p. 315). Asexually reproducing organisms are both interactors and replicators. Sexually reproducing organisms, on the other hand, are interactors but not replicators; their genes are replicators. Sexually reproducing organisms nevertheless are the entities which interact, as wholes, with their environments in such a way that the replication of their genes is differential; they are thus the entities upon which selection directly acts.

Most importantly, for purposes of this discussion, many multitaxon biofilms growing on nonliving surfaces form highly structured, metabolically integrated units. Yet biofilms do not qualify as "reproducers" in even the extended sense in which multitaxon symbiotic consortia (holobionts) were characterized earlier as reproducers.

Despite lacking physical boundaries, mature multitaxon biofilms have complex, three-dimensional architectures, consisting of functionally distinct ecological niches (Kirchman 2012, Ch. 3). The surfaces of biofilms growing on submerged surfaces exposed to sunlight, for example, provide ideal environments for photosynthetic and aerobic microorganisms. The former produce oxygen and nutrients that can be used by the latter in their metabolic processes. Oxygen concentrations deep within these biofilms are typically low, providing anoxic niches for anaerobic microorganisms. Chemoautotrophic archaea produce organic compounds from inorganic materials (e.g., carbon dioxide and hydrogen gas) that can be consumed by organotrophic microorganisms.

Given the opportunity, most planktonic (free-living) unicellular microorganisms will collaborate to construct a biofilm (Dang and Lovell 2015; Davey and O'toole 2000). The process is by cell aggregation. It begins when a few planktonic microorganisms (often bacteria) attach to a surface such as a submerged rock, mineral crust, boat hull, or shower stall, to name just a few illustrations. Other planktonic microorganisms join them. The phenotypic characteristics (physiological and behavioral) of a planktonic microorganism change when it joins a biofilm, enabling it to perform specialized, community supporting functions. Microbial cells (bacteria, archaea, and microeukaryotes) secrete extracellular polymeric substances (EPSs) that help anchor them to the surface and to each other (Decho and Gutierrez 2017; Sheng et al. 2010). In a mature biofilm these polymers form a thick, nutrient rich, extracellular matrix, encasing the community's microbial inhabitants; indeed, the presence of EPSs is what distinguishes biofilms from other microbial communities. The EPS matrix facilitates chemical communication (quorum sensing) between microorganisms, helping them to coordinate activities benefiting the entire community. It also protects members of the community from predators and harmful environmental conditions, such as desiccation and toxic chemicals (e.g., detergents) (Lopez et al. 2010). In addition, the matrix traps nutrients and breaks them down with extracellular enzymes (secreted by members of the community), in effect, serving as a community-wide digestive system (Flemming and Wingender 2010). Finally, the matrix facilitates HGT among members of a biofilm, enhancing the capacity of the community, as a whole, to adapt quickly to deleterious environmental changes. In short, the entire multitaxon community functions metabolically as a highly integrated, dynamically self-sustaining and self-repairing, cohesive whole.

As characterized above, the metabolic integration of many multitaxon biofilms found on nonliving surfaces rivals that of host–microbiome complexes.[11] It is thus not surprising that some researchers (e.g., Ereshefsky and Pedroso 2015, 2016; Shapiro 1998) have characterized them as multicellular organisms. Biofilms are nonetheless difficult to theorize as reproducers. The bulk of the microbiota of a host–microbiome complex are transferred from a fairly small number of holobionts of the same kind (e.g., humans with whom they come in contact during and after birth). The microorganisms making up multitaxon biofilms, in contrast, may come from many different sources, some shed from biofilms of very different types and others descending from several generations of free-living cells. This makes it difficult to conceive of them as forming unified reproductive (parent–offspring) lineages; there are too many dissimilar "parents" making widely variable contributions to the taxonomic makeup of the biofilm. Furthermore, taxonomically distinct biofilms may be very similar biochemically, and biochemically similar but taxonomically different organisms can replace each other in "the same" (a persisting) biofilm. The latter suggests that the identity of a biofilm, as a biological individual, may be more fruitfully construed, for theoretical purposes, in terms of the functional roles of the microorganisms making it up than in terms of the taxa to which they belong (Burke et al. 2011). This is not to deny that the microorganisms in a biofilm reproduce and participate in natural selection as individuals. The issue at hand, however, is whether biofilms participate in the process of natural selection *as unified wholes.*

Marc Ereshefsky and Makmiller Pedroso (2015) sum up the challenge posed by highly structured, metabolically integrated biofilms as follows: ". . . if biofilms are evolutionary individuals, a clear choice may be made. Either some evolutionary individuals do not reproduce, or the notion of reproduction should be expanded to include at least some cases of aggregation" (p. 10129). Drawing upon Griesemer's (2000) novel account of reproduction, they defend the latter position (Ereshefsky and Pedroso 2015, 2016): Biofilms "reproduce" by a process of microbial cell aggregation and the cells involved may come from many different sources. Thus their account of the reproduction of a biofilm represents a further expansion of the traditional evolutionary concept of reproduction.

As discussed in Section 4.3 (and revisited in the discussion of species, in Section 6.2), it is always possible to salvage a venerable theoretical concept by reinterpreting it to accommodate problematic cases, and if such efforts fail to yield a universally applicable interpretation one can always fall back on a pluralist

---

[11] It is not too surprising that the metabolic unity of many biofilms is similar to that of holobionts. As mentioned earlier, the microbiota of holobionts form an extensive network of biofilms. What I wish to explore in this section is whether multitaxon biofilms on nonliving surfaces can be said to be reproducers in the sense required for participating directly in natural selection.

account of the concept. This is the strategy endorsed by Ereshefsky and Pedroso. They conclude that there is no universal mode of biological reproduction. The concept of reproduction is disjunctive, with some Darwinian individuals (e.g., viruses) satisfying a replicator account, like Hull's, others (e.g., animals) satisfying a bottleneck account, like Godfrey-Smith's, and still others (e.g., biofilms) falling under Griesemer's account. At a certain point, however, one wonders whether (as Section 4.4 discussed) the problem lies with the concepts and principles being defended come what may. The motivation for classifying host–microbiome complexes and certain multitaxon biofilms as biological individuals is based on their remarkable structural and metabolic (as opposed to evolutionary) unity. What is driving efforts to theorize them as reproducers is the assumption that biological individuals must be Darwinian individuals. But what if this assumption represents a eukaryote-centric bias? After all, the traditional account of biological individuality, in terms of unified reproductive (parent–offspring) lineages, is especially well suited for complex multicellular eukaryotes, especially animals, which is hardly an accident since it was designed with them in mind. It is clear, however, that the traditional account of Darwinian individuality cannot be extended to all known structurally and metabolically unified biological entities without rendering the key concept of reproduction almost unrecognizable. Could what we are learning about the microbial world be telling us that we need to either rethink the dependence of natural selection on reproduction or, even more radically, embrace the biological plausibility of living individuals that are not Darwinian individuals? The remainder of this section is devoted to some speculations along these lines.

Although they do not reproduce, biofilms do *regenerate* (achieve persistence by reconstructing themselves). Complex multicellular eukaryotes also regenerate, and although most of them reproduce as unified wholes, not all of them do, for example, mules and spayed cats. Yet few would deny that sterile animals are living individuals on the grounds that they cannot reproduce.[12] Furthermore, the ways in which biofilms and multicellular eukaryotes regenerate, primarily, by cloning (replicating) individual cellular components and incorporating them into the whole (biofilm or animal) in a piecemeal fashion, is similar; the cells of both mules and donkeys are replaced separately, as opposed to all at once. Reproduction in multicellular eukaryotes, in contrast, has a bottleneck-like character. Each generation is created anew from a single cell. Regeneration, not reproduction, is what allows both microbial consortia (biofilms and microbiomes) and (fertile and sterile)

---

[12] As discussed in Section 2.2.1 only when defending definitions of life are scientists and philosophers willing to bite the bullet and reject sterile organisms, such as mules and spayed cats, as living things. This strategy is very problematic, however, insofar as it is motivated solely by the desire to protect a preferred definition of life from what would otherwise be an intractable counterexample.

multicellular eukaryotes to sustain themselves structurally and metabolically as individuals (in a far from equilibrium state vis-à-vis their environment) for an otherwise improbable length of time. Indeed, without the capacity to regenerate, an animal would not survive long enough to reproduce.

Admittedly, there is a striking difference between the regeneration of a donkey or mule and that of a biofilm. The regeneration of the former is under the control of an overarching genetic program. The whole (mule or donkey) is continuously reconstructed through replication of individual somatic cells and the process is coordinated by their shared genome. Biofilms do not regenerate in this fashion. Their microbial inhabitants do not share a unique genome and, as a consequence, the regeneration of the biofilm is not coordinated at the level of cell replication. Instead, the organization of the biofilm is passed on to its microbial constituents through its pre-existing spatial structure, the three-dimensionally organized, specialized microenvironments found within it.[13] Anaerobic microorganisms cannot survive in oxygen rich microenvironments. Similarly, photosynthetic microorganisms cannot survive in places where sunlight never penetrates. In short, life as we know it uses two different regenerative strategies to achieve persistence of the whole through time, both of which differ from reproduction (in the traditional evolutionary sense). One organizes cell level replication by means of a shared genetic program whereas the other imposes organization on cell level replication through the three-dimensional spatial structure of the whole. Nevertheless, both modes of regeneration serve the same biological function, namely, sustaining the whole as a coherent structural and metabolic unit for an otherwise improbably long period of time. In this light, the provocative question arises: Is the replication of a prokaryote, by simple cell division, better theorized as a form of regeneration or reproduction?[14]

To the extent that regeneration is critical to the capacity of a biological system to persist, and persistence is required for evolution to occur, regeneration seems more basic to being a living thing than reproduction. Besides, the concept of a living individual unable to reproduce but nonetheless persisting through time by

---

[13] Regeneration in this sense resembles what Godfrey-Smith calls "scaffolded reproduction" but it does not involve generating a separate individual.

[14] The Ship of Theseus Paradox, familiar to every student of philosophy, suggests that regeneration and reproduction may be more closely related than suggested earlier. The dilemma is whether a ship whose planks are gradually replaced, one at a time, is the same ship when the process is finished. Part of the puzzle lies in the fact that if one built a separate ship next to the original ship from the very same planks there would seem to be two ships. But what if one destroyed the old ship, leaving the new ship by itself? Is there a principled difference between the first case and the last case? On both scenarios we end up with a single new ship built from the same planks. Regeneration, in the biological sense, closely resembles the piecemeal replacement of a ship in a plank by plank fashion, whereas reproduction, in the traditional biological sense, resembles the building of a second ship next to the original ship. The analogy is heightened when one considers that offspring typically outlive their parents. Viewed in this light, reproduction and regeneration seem to be closely related mechanisms for preserving biological individuality. Unfortunately, it is beyond the scope of this chapter to explore this interesting analogy further.

continuously renewing itself seems perfectly coherent. Yet the idea of a persisting individual that cannot reproduce poses a challenge to traditional evolutionary theory, in which genetic-based reproduction plays the central role. Is it possible to reconstrue evolutionary theory in terms of persistence, as opposed to reproduction?

In some intriguing papers, Ford Doolittle (e.g., Doolittle 2017; Doolittle and Booth 2017) proposes expanding the concept of natural selection to include "differential persistence" (in addition to differential reproduction). The basic idea is to treat biogeochemical cycles and other homeostatic mechanisms of microbial communities, as opposed to the collections of microorganisms, and their taxa (species, etc.), constituting them, as units of selection. For as Doolittle notes, functionally similar microbial communities may be composed of very different taxa, and the taxonomic makeup of a microbial community (whether holobiont or biofilm) may vary over the course of its lifetime and the lifetimes of its "descendants." What is conserved, from one generation to the next, is the biochemical functions of its members and their contributions to the community as a coherent whole. In Doolittle's colorful words, it is "the song, not the singers" that is transferred from one generation to the next. In this context, one might wonder whether reproduction represents a specialized adaptation of the complex eukaryote cell structure (with membrane enclosed organelles, such as mitochondria and a chromosome containing nucleus) to the extracellular environment.

It is beyond the scope of this section to do more than raise these vexing questions. How one answers them has important ramifications for the search for novel forms of life, however: Could there be planets or moons in our solar system with forms of life that never developed the capacity to reproduce but are nonetheless able to sustain themselves for long periods of time as individuals by continually reconstructing themselves in an adaptive manner? As the next subsection discusses, some biogeochemists suspect that the first living things on our own world may have been acellular geochemical systems of this sort. Could such systems still exist on Earth, unrecognized for what they represent?

As a final point, in virtue of representing paradigmatic examples of living things, animals (and to a lesser extent plants) are still driving biological theorizing. Instead of following Dupré and O'Malley in theorizing host–microbiome complexes as heterogenomic *multicellular* organisms, a more microbe-centered approach would be to theorize multicellular eukaryotes as highly specialized microbial communities. Put provocatively, could our elaborate, monogenomic, somatic-celled bodies be little more than evolved "space suits" for microbial communities – sophisticated adaptations allowing microbes to expand into environments that were once inaccessible to them? In this context, it is perhaps significant that (as mentioned earlier) true multicellular eukaryotes emerged less than a billion years ago, when the Earth's environment became more stable, allowing multicellular eukaryotes, which are far less resilient in the face of environmental fluctuations than prokaryotes, to thrive. Less than a billion years from now, when the sun begins its death throes, slowly expanding into a red giant, bacteria and archaea will still be found on Earth, long after true multicellular eukaryotes have ceased to exist.

6.3.3 Could Rock-Powered Ecosystems Be Living Things?

Ecosystems are communities whose components include both living things and nonliving things. As this subsection discusses, the nonliving components of some ecosystems are functionally integrated with their living components in an organism-like way, raising the question of whether they too should be classified as living individuals. The infamous Gaia hypothesis (Lovelock 1972; Lovelock and Margulis 1974), which holds that Earth's global ecological system is a living individual consisting of both living and nonliving components, is the most extreme proposal along these lines. It has been trenchantly criticized, however; see, for example, Tyrell (2013). A more compelling case can be made for some intriguing, localized, microbial ecosystems that are dynamically maintained in a stable biogeochemical state through tightly coordinated, mutually facilitated, interactions among certain of their microbial and nonliving components.

Biological metabolism is traditionally theorized as consisting of two subprocesses, *catabolic* reactions, which break down complex molecules into simpler ones, releasing energy in the process, and constructive *anabolic* reactions, which synthesize complex biomolecules (e.g., ATP, protein, and nucleic acids), required for maintenance, growth, and repair, from materials supplied by catabolism. In animals and plants, catabolism is usually internal and anabolism is always internal. Bacteria in many microbial communities engage in community-wide extracellular catabolic processes. Instead of ingesting organic material and breaking it down internally, these bacteria excrete enzymes into the extracellular environment and break it down externally, providing other members of the community with a pool of pre-digested nutrients. Extra-organismal catabolism is not unique to microbial communities. Spiders inject digestive enzymes into the bodies of their victims and then suck out the liquefied contents. Multicellular eukaryotes do not, however, engage in external anabolic processes, and until recently this was assumed to be true of archaea and bacteria as well.

The first stage of anabolism consists of the transfer of electrons from electron donors to electron acceptors via redox (coupled oxidation–reduction) reactions. In eukaryotes these electron transport chains (ETCs) occur within the membranes of cell organelles (mitochondria and chloroplasts). In prokaryotes, ETCs occur within cell membranes. The flow of electrons along ETCs generates electrochemical gradients, which power the synthesis of adenosine triphosphate (ATP). Known as the "energy currency" of the cell, ATP is the primary source of energy for most cellular functions, including the synthesis of nucleic acids and proteins.

In organotrophs, such as animals, ETCs are fueled by catabolism (aka digestion) – the breakdown of complex organic molecules (carbohydrates), which releases electrons. The flow of electrons within membranes generates proton pumps (electrochemical gradients) that fuel the production of ATP. Some bacteria, for example, varieties of *Shewanella* and *Geobacter*, skip the catabolic stage of metabolism by syphoning electrons directly from minerals in rocks and transporting them to cell membranes via *extracellular* ETCs. In essence, these "electricity eating" bacteria go directly to the anabolic (constructive) stage of metabolism. Moreover, some of these bacteria also deposit electrons directly onto (electron-accepting) minerals in rocks, in effect "breathing electrons" and completing the redox cycle outside of a cell membrane; see Glasser et al. (2017) for a review.

How do electricity eating and breathing bacteria transfer electrons to and from minerals in rocks? The release of electrons is facilitated through water–rock reactions such as serpentinization, the aqueous alteration of olivine to serpentine. Bacteria extend hair-like filaments ("nanowires") to the surface of the rocks (El-Naggar et al. 2010; Sure et al. 2016). These filaments function as bioelectric wires, carrying electric charge from rock to cell in what amounts to extracellular ETCs. Bacteria isolated from these ecosystems have been kept alive by growing them directly on electrodes with no additional sources of nutrition (Summers et al. 2013); they are literally eating electricity. In some water–rock ecosystems, bacteria are interconnected by a web of biological nanowires (Figure 6.3), transporting energy, obtained directly from rocks, from cell to cell. Rock-powered ecosystems have been discovered in which bacteria line up end-to-end, forming living bioelectric cables stretching for centimeters (perhaps even longer), along which electrons flow to and from rocks and cells; see El-Naggar and Finkel (2013) for a review. The entire aquatic ecosystem (minerals, rocks, water, and bacteria) is connected together by what amounts to an electrical circulatory system.

The adaptive advantages to an aquatic microbial community of using extracellular ETCs can be substantial. Bacteria living in anaerobic environments (e.g., the bottom of a water column) can export electrons to regions of the community containing more oxygen (closer to the surface) or, barring that, transfer them directly to electron-accepting (oxidizing) minerals in rocks. Bacteria living in nutrient poor environments, on the other hand, can acquire electrons directly from minerals in rocks and transport them via extracellular ETCs to parts of the community not in contact with rocks, providing a much needed source of metabolic energy, or alternatively import electrons from more nutrient rich parts of the

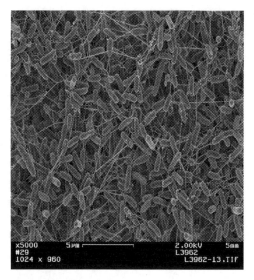

Figure 6.3 Photo of *Shewanella oneidensis* bacteria interconnected by nanowires. (Photo courtesy of Kenneth Nealson.)

community. A microbial community in a nutrient poor, anaerobic environment could, in principle, fully power itself metabolically by syphoning electrons released from minerals during water–rock reactions, cycling them among members of the community, and completing the battery-like extracellular circuit by depositing them back onto minerals in rocks. The point is some rock-powered ecosystems form electrical networks of remarkable complexity linking microbial cells to each other and to nonliving parts of the environment, raising the provocative question of whether the entire ecosystem – consisting of abiotic materials (minerals, rocks, and water), geochemical processes (water–rock reactions, such as serpentinization), biological entities (bacteria), and biochemical processes – is metabolizing as a coherent whole. Would it be more fruitful for biological theory to conceptualize these highly integrated, complex ecosystems as living individuals in their own right?

The extent to which rock-powered ecosystems may be characterized as engaging in full-blown, system-wide, metabolic processes is unclear. Extracellular ETCs do not drive the extracellular synthesis of ATP, let alone the synthesis of proteins and nucleic acids; these processes occur within bacterial cells. On the other hand, extracellular ETCs drive the proton pumps within cell membranes which power the synthesis of ATP. Moreover, bacteria in these ecosystems secrete chemicals (e.g., phenazines, flavins, and quinones) into the extracellular environment that can reduce or oxidize minerals, facilitating the release and deposit of electrons from and to rocks (Glasser et al. 2017). Some bacteria also generate

ETCs using organic matter (Watamabe and Nishio 2010). The upshot is that the living and nonliving parts of these ecosystems are functioning together to dynamically maintain the system, as a whole, in a stable state. Considerations such as these raise the question of whether the ability to synthesize particular molecules – ATP, proteins, and nucleic acids (or their functional equivalents) – is essential to biological metabolism.

At the core of the biological concept of metabolism is the notion of homeostasis. Homeostasis is usually characterized as the maintenance of an *internal* state of equilibrium in the face of degrading *external* conditions. In the case of rock-powered ecosystems there is no clear-cut distinction between inside and outside. The system is (in the terminology of philosophy) a 'scattered object', that is, a discontinuous whole whose components are interspersed with things that are not proper parts of it. The functional integrity of a rock-powered ecosystem, as a dynamically self-sustaining whole, is maintained through carefully orchestrated interactions between its living and nonliving parts that strategically move energy around the system to where it is needed. Even more provocatively, some rock-powered ecosystems appear to be very old, persisting for perhaps thousands of years. Why must the biogeochemical reaction networks keeping these ecosystems in (otherwise) improbably stable states of equilibrium for long periods of time be isolated from their environments by physically confining barriers in order for them to qualify as fully metabolizing in a biological sense?

Many origins of life researchers contend that enclosure by a selectively permeable membrane is required for isolating and concentrating the molecular building blocks of life (amino acids, cofactors, sugars, nucleotide bases, etc.), so that they can chemically react, and for protecting the resultant proto-biomolecules (peptides and/or RNA oligomers) from degrading chemical reactions; see Section 5.4 for a discussion. As a consequence, the origin of life is usually associated with the emergence of the first protocell(s), and the debate swirls around the question of whether they enclosed a primitive metabolic or genetic system (or both). What we are learning about rock-powered ecosystems suggests, however, that the first living things may not have been cells – cellular life could have been preceded by open-ended geochemical reaction networks powered by natural proton and redox gradients across thin mineral layers found in deep sea, alkaline hydrothermal vents (Koonin and Martin 2005; Lane and Martin 2012; Martin and Russell 2003).[15] In the words of Lane and Martin (2012),

---

[15] The conjecture that the first living things may not have been enclosed in vesicles is actually older – see, e.g., Cairns-Smith (e.g., 1982), Dyson (1999), and Wachterhauser (e.g., 1992) – but what we have learned in recent years about rock-powered ecosystems has allowed researchers to flesh out more detailed geochemical scenarios as to how this could have happened.

... natural proton gradients can, in principle, drive the beginnings of an anabolic biochemistry [driving the production and accumulation of amino acids, bases, sugars and lipids], eventually forming protocells within the vent pores... In early stages, we envisage networks of inorganic compartments lined distally to the ocean with leaky organic membranes but proximately contiguous with vent effluent... But to escape the vents as independent free-living cells requires a switch from relying on natural proton gradients to forming true cells capable of actively generating ion gradients on their own.

(p. 1411)

Lane and Martin's proposal for rethinking the origin of life raises an intriguing question: Is cellularity truly universal to life, or merely contingent to life found in certain types of environments, such as on the early Earth? Could there be forms of life which never give rise to free-living, self-replicating cells? How one answers these questions has consequences. The most popular theories of the origin of life (RNA World and SM World) assume that the first living things were encapsulated in organic membranes, and as discussed in Section 5.4, these models are plagued with difficulties. Could the assumption that life began with protocells be hindering the development of a more satisfactory model of the origin(s) of life? Similarly, most astrobiologists are looking for cells or their (chemical or structural) remnants. The danger is that they are (so to speak) looking under the streetlight and will fail to recognize an alternative form of life should they be fortunate enough to encounter it elsewhere in our solar system.

6.4 Concluding Thoughts

This chapter raises many questions, some more farfetched than others from a contemporary biological perspective. The goal is not to answer them. It is to be provocative – to draw attention to the potential significance for biological theory of some of the more striking ways in which the microbial world is challenging the neo-Aristotelean, eukaryote-centric, model of life. For as discussed in Chapter 4 (and further developed in Section 8.3), discoveries that challenge our theoretical preconceptions provide the best grist for the theoretical mill, clearing a path to new and better theoretical frameworks for generalizing about a domain of empirical phenomena. The key is recognizing such phenomena for what they represent, instead of ignoring them or, in accordance with the fashionable pluralist stance, treating them as yet another nail in the coffin of universal biology.

As Chapter 4 discussed, what a fruitful program of universal biology needs is more unifying theoretical concepts – concepts applying to (unicellular and multi-cellular) eukaryotes, prokaryotes, and acellular entities, such as viruses – for sorting known Earth life into categories that are better at revealing truly general patterns in the origin, development, and diversification of life on Earth, and hopefully elsewhere in the universe as well. Unfortunately, as the history of science reveals (Sections 4.2–4.4), scientists tend to retain well entrenched theoretical concepts even when the principles in which they figure face exceptions which have resisted persistent efforts to resolve them. It is thus not surprising that some biologists and many philosophers of biology are pluralists about key biological concepts, such as that of species, reproduction, and living individual, contending that there are a variety of incommensurable but equally legitimate ways of construing them. As Section 4.5 discussed, however, the trouble with taking a pluralist stance towards such concepts is that it undermines the development of more fruitful classificatory schemes for generalizing about the phenomena concerned. Indeed, perhaps – as was the case with the now defunct concepts of phlogiston and the luminiferous aether (discussed in Section 4.4) – the problem lies in the very concept of species or living individual. One is very unlikely to recognize unity under the assumption that the phenomena concerned are intrinsically heterogeneous. In contrast, it is much easier to recognize disunity (the manifestation of problematic exceptions) under an assumption of unity. It is for this reason that I recommended a monist over a pluralist stance in Section 4.5.

There are, of course, no guarantees that researchers will discover truly unifying concepts and principles for life from investigating the microbial world. It is possible that pluralists are correct about life lacking an underlying unity. It is also possible that life on Earth is unrepresentative or that our theorizing about life is being held hostage to characteristics of Earth life that (unbeknownst to us) are contingent upon physical and chemical conditions on the early Earth. Nevertheless, it seems clear that seeking basic concepts and principles from the microbial world is more likely to yield a successful foundation for universal biology than seeking them from multicellular eukaryotes, which we now know are outliers (highly specialized, fragile, latecomers) to our planet. At the very least, such a search will provide us with a more comprehensive understanding of life as we know it on Earth today. But, again, to achieve such an understanding we need to stop trying to cram the microbial world into a theoretical framework based on multicellular eukaryotes.

On the other hand, it is clear that what is really needed for formulating a truly universal theory of life are examples of life as we don't know it. The remaining chapters in this book investigate three strategies for acquiring such examples. Chapter 7 explores the question of whether the creations of artificial life hold forth the promise of providing us with truly alternative examples of life. Chapter 8 discusses how to search for novel forms of extraterrestrial life in the absence of a definition or universal theory of life. Last but certainly not least, Chapter 9 investigates the intriguing possibility that Earth itself may be host to as yet undiscovered alternative forms of microbial life.

7

Artificial Life: Could ALife Solve the $N = 1$ Problem?

7.1 Overview

Some artificial life (ALife) researchers contend that we are on the verge of creating novel life forms either in the form of information structures in a computer (soft ALife), robots made of plastic and metal (hard ALife), or synthetic organisms composed of unnatural biomolecules (wet ALife or synthetic biology). In the words of computer scientist Chris Langton, a founder of ALife, "Artificial life can contribute to theoretical biology by locating life-as-we-know-it within the larger picture of life-as-it-could-be" (Langton 1989, 1). Similarly, in a discussion of synthetic life, Mark Bedau and colleagues note that "[o]ne of artificial life's key goals is constructing a life form in the laboratory from scratch" (Bedau et al. 2000, 365). This chapter evaluates whether ALife research can live up to its hype and deliver truly novel forms of life. As will become apparent, the inventions of ALife are very closely based on characteristics of familiar Earth life. In addition, they tend to fall outside of the Goldilocks level of abstraction (Section 4.3) in being either too abstract (soft and hard ALife) or too concrete (wet ALife). It is thus highly unlikely that ALife can help us solve the $N = 1$ problem, which is not to claim that it cannot teach us important things about life.

7.2 Soft ALife: Digital Organisms?

Our concern in this chapter is with *strong* versions of ALife, that is, attempts to literally create novel living entities. Strong soft ALife and hard ALife are predicated on the assumption that life is an abstract functional process independent of its material composition. Any system exhibiting the pertinent abstract functional properties, whether a computer simulation (soft ALife) or a robot composed of plastic and metal (hard ALife), qualifies as a living entity. In Langton's words, "the principal assumption made in [strong hard and soft] Artificial Life is that the 'logical form' of an organism can be separated from its material basis of construction, and that 'aliveness' will be found to be a property of the former, not of the latter" (Langton 1989, 11). The "logical form" of life is invariably identified with either an algorithmic rendering of Darwinian evolution by natural selection or a formal analysis of metabolism, most commonly as autopoiesis. For advocates of the former, any dynamic pattern satisfying a favorite Darwinian algorithm – at minimum, exhibiting the transmission of heritable small variations and evolving according to a generalized principle of natural selection – is *ipso facto* a living thing. For advocates of the latter, any dynamic system exhibiting autopoiesis, which generalizes the capacity of biological organisms to sustain themselves as complex self-organized systems through regenerating their components, is a living entity (Luisi 1998; Maturana and Varela 1973; Zeleny 1981). According to Langton and fellow travelers, the composition of a system exhibiting the privileged logical form (Darwinian evolution or autopoiesis) is irrelevant to its status as a *bona fide* living thing.

The goal of most soft and hard ALife research is not creating life but rather achieving a better understanding of the genetic, metabolic, and behavioral processes of life by simulating them on computers or modeling them with robots.[1] There are some notable exceptions, however. Ecologist Thomas Ray is perhaps the best-known advocate of strong soft ALife.[2] In the early 1990s, Ray developed a computer simulation, dubbed "Tierra," in which computer programs compete for time on the CPU (central processing unit) and space (access to main memory). Ray (1994a, 1994b) argues that these programs evolve via Darwinian evolution in the same manner as biological organisms, mutating, self-replicating, and competing, and hence that they qualify as authentic living individuals. According to Ray, Tierra's digital organisms are more alien than any extraterrestrial organisms that astrobiologists may someday encounter (Ray 1991).

How plausible are Ray's bold claims? As he freely admits, the claim that Tierra's programs are living entities is grounded in the assumption that being a living thing is merely a matter of satisfying "the" Darwinian algorithm, regardless of how this is accomplished (Ray 2009, 429). But even supposing that Tierra's programs satisfy Ray's version of the Darwinian algorithm, questions have been

---

[1] The distinction between this weaker form of ALife and strong ALife is analogous to a famous distinction between weak and strong AI (Artificial Intelligence). Strong AI aims at creating machines that not only exhibit intelligence but also have minds – are conscious (self-aware) and sentient (capable of feeling pain). Weak AI, in contrast, is concerned only with achieving a better understanding of (human) cognitive processes through simulating them on computers.

[2] Elon Musk, CEO of SpaceX and Tesla Motors, took an even more extreme position at Code Conference 2016 when he opined that humans are most likely simulated creatures living in a simulated reality created by a super intelligence. Musk's stance raises the question of the existential status of the super intelligent beings that reputedly programmed us into existence. Is it turtles all the way down (or is it up?)? In any case, the tendency to conflate simulation with reality is becoming increasingly popular but remains deeply problematic.

raised about whether the latter captures evolution in the biological sense. Biological systems on Earth exhibit "open-ended evolution," a process by which they continue indefinitely to produce new forms. Tierra and other digital evolutionary systems do so for a time but eventually reach a state in which they generate only forms that were previously produced, in essence ceasing to "evolve" (Bedau et al. 2000; Standish 2003). Open-ended evolution has thus far eluded capture in a formal algorithm that can be implemented on a computer. Doing so remains a central goal of soft ALife research, however, and efforts to algorithmatize openended evolution are ongoing (e.g., Soros and Stanley 2014).

Most ALife researchers believe that open-ended evolution will eventually be realized on a computer. If this happens, have they thereby created digital life? The answer is not as straightforward as Ray and like-minded researchers suggest. An affirmative answer presupposes not only that a functional characteristic, very closely based on familiar Earth life (Darwinian evolution that is open ended), is all that matters to being a living thing but also that a very high level of abstraction (as captured in an algorithm) is the correct one for generalizing about that function. Proponents of metabolic accounts of life would not even get past the first stage of this reasoning. Their favorite functional characteristic of familiar life is metabolism.

Margaret Boden's (1996, 2000) discussion of biological metabolism, which includes a critique of Ray's Tierra, is especially salient for purposes of this chapter. She argues that metabolism is basic to life and that metabolism in the biological sense is not just a matter of dynamical organizational complexity but of *self-organization*, which involves "... the use, and budgeting, of energy for bodily construction and maintenance (and behavior) ..." (Boden 2000, 121). According to Boden, metabolism in this sense cannot be captured in a formal algorithm and implemented on a computer. Boden explicitly attacks the claim that autopoiesis – the favored abstract version of metabolism for soft ALife researchers (Agmon et al. 2016)[3] – could be fully captured in a computer simulation, although efforts to do so continue (e.g., Cornish-Bowden et al. 2013).

The view that metabolism cannot be realized by a computer simulation is fairly common. Indeed, some researchers contend that the deeply self-organizing character of metabolism, which is alleged by some to be captured in the abstract concept of autopoiesis, is analogous to the self-referentiality underlying the infamous halting problem of computability theory (Kauffman 1993; Rosen 1991). While it is beyond the scope of this chapter to develop the analogy, which would require explaining the highly technical halting problem, it suffices to note that *if* biological

---

[3] Autopoiesis is discussed in Chapter 2 (Section 2.2.4). While there are different versions, the core idea is a self-contained system organized as a network of processes of production and destruction of components that continuously regenerates itself (Maturana and Varela 1973).

metabolism is self-referential in this sense, then it is Turing uncomputable, which means that it cannot be implemented on a digital computer. One of the more interesting aspects of this debate is that it resurrects the old question, discussed in Chapter 1, of whether life uses a strange form of causation that cannot be "reduced" to the linear form of causation familiar from classical physics and presupposed in the Turing machine conception of computability; I have argued elsewhere that the Turing machine conception of physical computability is not nearly as compelling as often supposed (Cleland 1993, 2004, 2006).

Even supposing, for the sake of argument, that a computer simulation were able to realize autopoiesis, there are three problems with the claim that it would thereby qualify as a living thing. (i) A particular biofunction (metabolism) is being favored as fundamental to life over others (most notably, open-ended evolution); (ii) a very Earth-centric version of metabolism is being privileged over others as fundamentally biological; (iii) a very high level of abstraction (a formal version of autopoiesis) is being privileged as the correct level for generalizing about biological metabolism. As adumbrated earlier, analogous worries hold for anyone arguing that a computer simulation of open-ended Darwinian evolution is a novel living thing. Moreover, these difficulties cannot be overcome by jointly realizing metabolism (in the favored form of autopoiesis) and open-ended Darwinian evolution in a computer simulation. For even supposing that metabolism and evolution are fundamental to life, the question of whether the highly abstract versions realized in the simulation are the right ones for generalizing about them remains.

7.3 Hard ALife: Living Robots?

In light of the considerations canvassed above, perhaps hard or physically embodied ALife provides a more promising potential source of novel life forms. Like strong soft ALife, strong hard ALife holds that being alive is an abstract functional property of a system. According to the latter, however, this life conferring property must be physically realized via causally interacting components. The "parts" of a digital organism, in contrast, are merely logically interrelated, despite the fact that the simulation is implemented on a physical machine (computer) whose components (electronic circuits, etc.) do causally interact. The "legs" of a concrete robot (hard ALife creature) are partially identifiable with its physical components in a way that is not true of the parts of a digital organism. Although they must be physical, the components of a hard ALife construction need not be chemical, let alone carbon based. Robots composed of plastic parts and metal wires will qualify as living things if they causally realize the pertinent abstract functional properties *in the right way.*

One of the most famous proponents of strong hard ALife, Hod Lipson, identifies life with the capacity to evolve by natural selection and self-reproduce (e.g., Lipson and Pollock 2000; Zykov et al. 2005). Lipson's robots are created in two stages. First, an evolutionary algorithm is used to "evolve" a simulation of a robot in a virtual world on a computer. A 3D printer subsequently turns the simulated robot into a solid, three-dimensional object. In essence, the evolutionary algorithm generates a blueprint for a robot and the printer then constructs it. The process is not fully automated. A human being needs to wire the robot and put in a battery, among other things. Lipson's goal is eventually to automate the entire process, that is, to write a sophisticated computer program capable of evolving a simulated robot (adapted to a particular environment in the natural world) and then to construct it by controlling the appropriate machinery. According to Lipson, the result will be a truly novel form of life – an evolved, self-reproducing (and therefore living) robot composed of inorganic material.

Lipson's ambitious project faces problems, some more serious than others. First there are questions about whether Lipson has identified the appropriate level of abstraction for generalizing about self-reproduction and Darwinian evolution. In the biological realm, evolution cannot occur without physical replication, and hence both are limited by the causal (versus logical) constraints of the material (versus a simulated) world. There is also Boden's concern about metabolism. Biological organisms do not construct their physical bodies and then power them by snapping in a prefabricated energy source. They rely on their "batteries" (electron transport chains, proton pumps, and molecules of ATP) from the beginning to power the process of self-assembly.

These difficulties bring us to the question of what, on Lipson's account, could constitute a living robot? His reply seems to be: the physical robot. But surely this response cannot be right. The physical robot is the end product of an external, protracted manufacturing process. It does not assemble itself. It is constructed by a 3D printer under the control of a computer program following a blueprint evolved on a computer. Viewed in this light, the most appropriate candidate for a living Lipsonian robot is the factory-like extended manufacturing system consisting of computer, 3D printer, and robot. But this claim is also problematic because neither the computer nor the 3D printer assembles itself. The only physical object that is assembled by the extended system is the robot and, to reiterate, it does not assemble itself. In the biological case, in contrast, the manufacturing process is internal to and tightly integrated with the material body being constructed *while* it is being constructed. The latter (not the former) is what true self-assembly is like.

Rasmussen and colleagues (2003) recognize the difficulties in getting a physical system composed of "geomaterials" to self-assemble in the biological sense and are working to overcome them. As with strong soft ALife, however, even if they succeed, the claim that the resultant devices (or, as they dub them, "proto-organisms") are living individuals rests upon the assumption that being a living thing is merely a matter of exhibiting a particular functional property, modeled very closely on familiar Earth life, at a high level of abstraction. As discussed below, there are serious difficulties with this claim.

The assumption that "aliveness" is an abstract functional property is deeply problematic. Functional properties that are difficult (and, in some cases, even impossible) to realize in a material system are easy to simulate on a computer using logical relations among logically primitive (not further analyzable) components. All one needs to do is pick a level of abstraction sufficiently removed from the causal constraints of the material world. The material composition of a physical system, however, is crucial to its causal capacity to perform a given function. Glass filaments will not conduct electricity even though one can easily design a computer simulation of glass filaments conducting simulated electricity; it is just a matter of picking an appropriate level of abstraction. Moreover, not just any arrangement of material components capable of causally realizing a function will do so. Bridges built from stones must be structured very differently from bridges constructed of wood if they are to be able to support the weight of an automobile. The point is genuine scientific understanding of a natural phenomenon requires paying close attention to causal interactions among the material entities involved. Rising to a level of abstraction where the latter considerations no longer obtain does not provide such understanding.

Modern science excels at explaining natural phenomena (physical objects, functions, properties, and processes) in terms of their detailed material compositions and natural (versus logical) "laws." The distinctive functional properties of water, for example, being a good solvent, expanding when frozen, and remaining a liquid over an unusually wide range of temperatures, are explained in modern chemistry in terms of water being composed primarily of H_2O molecules. Oxygen attracts electrons more strongly than hydrogen, producing a net positive charge on the hydrogen atoms and a net negative charge on the oxygen atom, turning H_2O molecules into electric dipoles. As a consequence, water molecules form weak hydrogen bonds with each other and tend to stick together, forming secondary structures. Secondary structures explain many of the unusual properties of water, such as remaining a liquid over a wide range of temperatures and expanding when frozen. These properties of water can be readily simulated on a computer but there is a critical difference between a simulated bottle of water blowing up in a simulated refrigerator and the physically real bursting of a bottle of water in an actual refrigerator. Understanding how a computer program generates the simulation tells us nothing about the natural processes causing a concrete bottle of water to rupture in a refrigerator; a clever computer scientist who knows nothing about the laws of chemistry or molecular theory can nonetheless write a program for simulating the phenomenon. The distinctive functional properties of other chemical substances are similarly explicable in terms of causal interactions among their chemical components. A computer simulation of a solution of hydrochloric acid will not dissolve the rust on my iron skillet, even supposing that it removes simulated rust from a simulated iron skillet.

There is compelling evidence that life (like water and rust) is a physicochemical phenomenon. As Section 5.2 discussed, familiar Earth life is composed of chemically distinctive molecules, most notably proteins and nucleic acids. Proteins and nucleic acids are unique to life as we know it on Earth; no known nonliving natural processes produce them. One of the great puzzles of biology is figuring out how their molecular precursors (peptides, nucleosides, and nucleotides) could be synthesized by abiotic processes under natural conditions. Proteins and nucleic acids causally realize the putatively basic functions of natural Earth life (metabolism and genetic-based reproduction). A simulated ribosome cannot translate an mRNA molecule into a biologically functional protein, nor can a simulation of photosynthesis produce edible sugars. It is much easier to simulate a drug destroying lung cancer tumors than to design a pharmaceutical agent that cures lung cancer. The point is strong versions of both hard and soft ALife presuppose an answer to the very question at issue: Is life, like water, a concrete thing comprising particular chemicals, or is it, like a song, an abstract process that can be hummed by a human being, played on instruments made of many different materials (from wood to metal to bone), and of course generated on a computer? In opting for the latter, defenders of strong soft and hard ALife ignore a large body of physical and chemical evidence that it is more like the former.

7.4 Synthetic Biology: Creating Novel Life in the Laboratory?

Synthetic biology (wet ALife) is our best hope for artificially creating a novel form of life. Biologically functional cells will eventually be created from scratch in the laboratory, and when they are, there will be little controversy about whether they qualify as *bona fide* living things. Indeed, Craig Venter and colleagues have already been credited with creating the first synthetic organism. They digitized (copied) the genome of a *Mycoplasma mycoides* bacterium and then synthesized and assembled it in the laboratory from basic molecular building blocks. When the synthetic genome was transplanted into a different variety of mycoplasma bacteria (*M. capricolum*) it exhibited the phenotypic characteristics of *M. mycoides* bacteria and replicated normally (Gibson et al. 2010). Several years later, Venter's team created a "minimal" bacterial cell containing just 473 genes (Hutchison et al. 2016). They first designed a new bacterial genome by removing genes that are not required for life from *M. mycoides*. The basic idea was to retain only genes that are essential for growth and self-replication. The team then synthesized and assembled the new genome in the laboratory and transplanted it into a bacterial cell whose genome had been removed. It took them a while to hit upon the right combination of genes, but eventually they produced a cell that could grow and self-replicate. Venter claims to have created a brand new species of bacteria, which he nicknamed "Synthia 3.0."

The claim that Synthia 3.0 is an artificially created organism is exaggerated, however. Venter's team inserted its artificial genome into a receptive cell whose genome had been removed but was otherwise functional; they did not synthesize the host cell's membranes, RNA molecules, or ribosomes from bottles of chemicals. Furthermore, while it is reasonable to characterize Synthia 3.0 as a new variety of familiar bacteria, it cannot be described as a novel form of life. Its genome uses the four standard nucleic acids (A, G, C, and T) and the same genetic code, and exploits the molecular machinery of familiar life for transcribing and translating information stored in DNA into proteins, which in turn are composed of biotic (standard) amino acids. In order to create alien (truly novel) microbes one needs "unnatural" biomolecules, composed of (at least) some nonstandard molecular building blocks (e.g., amino acids or nucleobases), something that Venter and his team have yet to accomplish.

Scientists are working on synthesizing novel microbial cells but progress is slow. That their progress is slow is not very surprising. Nucleobase pairs not occurring in nature have been synthesized in the laboratory and used to form unnatural DNA sequences (Sefah et al. 2014). But it is difficult to create functional microbes with the latter because most of them are incompatible with the transcription and translation machinery of familiar life. There is a notable exception. In 2012 a team of scientists at Scripps Research Institute in San Diego designed and synthesized a nonstandard base pair (known as d5SICS and DNaM) compatible with the enzymatic machinery that copies and translates DNA in normal cells (Malyshev et al. 2012). Two years later, they managed to create a plasmid (stretch of circular DNA) containing this nonstandard base pair (along with the standard base pairs, T–A and C–G) and insert it into an *E. coli* bacterium. The bacterial cell reproduced normally, copying and passing on the new DNA plasmid to daughter cells over multiple generations (Malyshev et al. 2014). Unfortunately, the new genes do not "do" anything yet, underscoring just how difficult it is to create alien microbes in the laboratory. The biomolecules of familiar life are tightly integrated at the molecular level in ways that are critical to their biofunctionality. Transcribing and translating even modestly unnatural DNA into biologically functional, unnatural proteins requires appropriate modifications to the core genetic machinery and metabolic pathways of familiar life. There is also the problem of enclosing the molecular processes of life within appropriately functional membranes. A team of chemists has created self-assembling lipid membranes from scratch by joining chains of lipids together using a simple metal ion (versus enzymes) as a catalyst (Budin and Devaraj 2012). Artificial genomes may someday be transplanted into vesicles enclosed by such membranes. But while undoubtedly qualifying as living things, the resultant cells are unlikely to help us much with the $N = 1$ problem. Cell membranes do not just separate the interior of a cell from its environment. They control the movement of chemicals into and out of cells, excluding toxins, admitting nutrients, and excreting waste products. Biologically functional, unnatural biomolecules, constructed from nonstandard molecular building blocks, will almost certainly require membranes that are selectively permeable to different varieties of molecules and ions; what are toxins and nutrients for us are unlikely to be toxins and nutrients for them. The point is creating viable alien microbes requires much more than synthesizing an artificial genome whose DNA contains some nonstandard amino acids. It requires tailoring nonstandard biomolecules to interact with each other in just the right ways to create an encapsulated chemical network exhibiting functions held up as fundamental to life.

Even supposing, however, that scientists were able to synthesize a biologically functional microbe from a suite of nonstandard lipids, proteins, and nucleic acids, it would still be closely modeled on familiar life. It is possible, of course, that all life is very similar in its molecular composition and architecture to our form of life, in which case life, like water and other chemical substances, is a distinctive chemical phenomenon. For all we know, however, life could be a higher-level physicochemical phenomenon, like a metal. Metals (iron, copper, aluminum, etc.) are distinguished from other material substances in terms of a suite of higher-level physical properties – being hard, shiny, a good conductor of electricity, etc. – that are nonetheless explicable in terms of lower-level chemical properties, such as losing valence electrons easily and being good reducing agents. It is even possible that life, like jade, is not a *bona fide* natural category. With the advent of molecular theory, minerals once classified as jade, on the basis of macroscopic physical properties such as color, hardness, crystal habit, etc., were found to consist of two distinct molecular compounds, now known as jadeite and nephrite. The point is we cannot infer the chemical and physical possibilities for life on the basis of a single example of life, especially one that we have reason to believe could be unrepresentative. The top-down creations of synthetic biology are unlikely to shed much light on this crucial issue. They will be created in our own image (so to speak).

On the other hand, if scientists synthesized a cell sharing some of the reputedly distinctive functional characteristics of life from alternative molecular compounds – compounds that are not merely modest variations on lipids, nucleic acids, and proteins – then there is little doubt that it would qualify as a truly alien form of life. The likelihood of such a project succeeding, however, seems extremely low. As discussed in Section 5.2, the biomolecules of familiar life are very large and complex. It is unlikely that a chemist who had never "seen" DNA would think it up or stumble upon it in the laboratory while experimenting with chemical reactions. Indeed, although they can synthesize them in a laboratory, biochemists still do not understand how DNA or RNA could arise from abiotic chemical reactions on the early Earth. Viewed from this perspective, it seems unlikely that chemists will be able to identify a suite of plausible alternative molecular compounds (most likely complex polymers of some sort) for an alternative biochemistry. For without guidance from a chemically based, universal theory of life, it is difficult to constrain the molecular search space except by trial and error, which seems unlikely to be fruitful given the number of possibilities that must be considered, both separately and, most importantly, combined, for realizing the highly integrated, basic functions of familiar life.

Some synthetic biologists hope to solve this problem by using machine-learning algorithms to stochastically explore molecular space for the purpose of predicting which molecular compounds are likely to produce key biological functions (e.g., Caschera et al. 2011). The basic idea is to use these predictions to winnow the molecular possibilities and focus experimental work on alternative biochemistries that are likely to yield promising results. Such investigations just might yield a better understanding of the chemical possibilities for realizing the biofunctions of familiar life. Still, the size and complexity of the molecular space to be explored is enormous, and there is no guarantee that a seemingly promising compound found in molecular space will be capable of participating with other molecular compounds in building a highly complex, functionally integrated, biochemical system of the sort thought to be required for life. Last but certainly not least, the question of the extent to which our current, Earth-centric, understanding of biological metabolism and evolution are generalizable to all forms of life remains.

7.5 Concluding Thoughts

ALife is unlikely to help us overcome the infamous $N = 1$ problem of biology. For as this chapter argues, the inventions of ALife researchers are modeled too closely upon either high level, functional characteristics of familiar life or low level, concrete (molecular) characteristics of familiar life. There are compelling scientific reasons for thinking that (i) life is more like water than a song, and (ii) that life could have been at least modestly different at the molecular level (Section 5.2).

What we really need are examples of natural life descended from more than one abiogenesis. Chapters 8 and 9 address this problem. Chapter 8 explores a novel strategy for acquiring such organisms without guidance from a definition or universal theory of life. Chapter 9 explores the possibility of a shadow biosphere, a form of microbial Earth life descended from an alternative origin of life. As will become clear, there is little theoretical or empirical support for the widely held assumption that such organisms do not exist, and moreover the tools used to explore the microbial world are too closely based on familiar life to detect them.

8

Searching for Extraterrestrial Life Without a Definition or Universal Theory of Life

8.1 Overview

When most people think about extraterrestrial life they envision intelligent, often technologically advanced, humanoid creatures, such as the Prawns (*District Nine*) and the Na'vi (*Avatar*), robots, such as Gort (*The Day the Earth Stood Still*), and the hive-like Borg (*Star Trek*). Most astrobiologists, however, are not looking for intelligent life. They are looking for bacteria-like organisms. For as discussed in Section 5.2, microbial life is almost certainly far more common in the universe than complex multicellular organisms, let alone intelligent, technologically sophisticated creatures. Because they are so tiny, detecting and identifying extraterrestrial microbes is especially challenging.

8.2 A Case Study: The Viking Missions to Mars

Just how difficult it is to search for alien microbes on the basis of our knowledge of familiar Earth life is underscored by the notoriously ambiguous results of NASA's Viking missions to Mars. The Viking missions consisted of two spacecraft, each comprising an orbiter and a lander. In the summer of 1976 the landers touched down in different regions of Mars. Each conducted three biology experiments designed to search for microbial metabolism; see Schuerger and Clark (2008) for a discussion of all three experiments. To this day the results of one of those experiments, the "Labeled Release" (LR) experiment, remains controversial, with the PIs (Principal Investigators) of the experiment (Gilbert Levin and Patricia Straat) insisting that they discovered Martian life and the majority of the scientific community disagreeing.

The LR experiment was based closely on the metabolism of familiar Earth microbes. Martian soil was deposited into a test chamber and inoculated with a dilute nutrient solution, radioactively labeled with ^{14}C, known to be metabolizable by a wide variety of bacteria that could be cultivated in the laboratory. The LR biology team anticipated that if Martian soil contained microbes, they would behave like bacteria and metabolize one or more of these organic compounds, releasing radioactive $^{14}CO_2$ gas in the process. To the team's delight this seemed to happen. $^{14}CO_2$ gas was immediately and continuously released. Indeed, the magnitude of the response was greater than occurred in prelaunch tests of the experiment on bacteria. Moreover, when a controlled experiment, involving heating Martian soil to a temperature high enough to kill any known microorganisms (160 °C for 3 hours) and then injecting it with the LR nutrient solution, was performed, $^{14}CO_2$ gas failed to be released. Levin and Straat were thrilled. They were certain that they had discovered Martian life (Levin 1997; Levin and Straat 1976, 1977, 1979).

Before announcing their discovery, the LR team decided to give a second injection of LR nutrients to samples of Martian soil that had evolved $^{14}CO_2$ gas in response to a first injection. They anticipated the release of more $^{14}CO_2$ gas as now hungry Martian microbes gobbled up more nutrients. This did not happen. Even more strangely, approximately 30% of the gas released during the first injection disappeared, apparently reabsorbed by the soil (Levin and Straat 1976, 1977). Not only was this inconsistent with the behavior of bacteria in samples of Earth soil, it was also difficult to explain in terms of abiological geochemical processes. In short, taken together, the results of the LR experiment were unexpected and ambiguous with respect to the question of whether they were the product of biology or nonliving processes. The initial response was strongly suggestive of microbial life but the failure to release $^{14}CO_2$ gas after a second injection of nutrients and the mysterious reabsorption of $^{14}CO_2$ gas produced after the first injection were difficult to explain either biologically or abiotically.

The way the scientific debate over how to interpret the anomalous results of the LR experiment unfolded is revealing. The PIs who played the lead role in designing the LR experiment insisted that they had found life on the grounds that the results technically satisfied (what amounted to) an agreed upon definition of life. In Levin's words, "[t]he Labeled Release (LR) life detection experiment aboard NASA's 1976 Viking mission reported results which met *the established criteria for the detection of living microbes* [italics are mine] in the soil of Mars" (Levin 1997). This illustrates the danger (also discussed in Section 2.2) of basing a search for extraterrestrial life on a definition of life. The results of the LR experiment were ambiguous for life. Levin's focus on whether they satisfied preconceived criteria for life obscures this important fact.

The leader of the Viking Biology Team concluded that they had not found life (Klein 1978a, 1991), and this remains NASA's official position (Beegle et al. 2007, 549). Various reasons have been proposed for rejecting the possibility of a biological explanation for the LR results. Bacteria typically reinvigorate when given a second injection of nutrients, which did not happen in the case of the LR experiments. Bacteria do not reabsorb carbon dioxide released during earlier metabolic processes when given a second helping of nutrients. The speed and magnitude of the positive response to the initial injection of LR nutrients was greater than that exhibited by known bacteria. In essence, these responses amount to opting for even more Earth-centric metabolic criteria for life than the one upon which the Viking experiments were originally predicated, which helps to explain Levin's and Straat's frustration.

The linchpin in the Viking Biology Team's case against Martian life, however, was the failure of the gas chromatograph mass spectrometer (GCMS) to find any organic compounds in Martian soil (Biemann et al. 1976, 1977). A miniature GCMS was included in the biology instrument package. It was not intended as part of the life detection equipment. Nonetheless, the failure of the GCMS to detect *any* organic molecules in Martian soil heated to 500 °C was cited by Klein and fellow travelers at NASA as the main reason for concluding that abiotic geochemical processes were responsible for the puzzling results of the LR experiment.

The results of the GCMS are problematic, however. Meteoritic dust falls continuously to the Martian surface and contains enough organics that the GCMS should have detected some. Even worse, research shows that the GCMS could not have detected as many as 10^6 microbes/gram of soil, more microbes than found in some samples of Antarctic soil (Glavin et al. 2001; Klein 1978; Navarro-González et al. 2006). In a nutshell, the Viking GCMS was not sensitive enough to pick up low levels of microbial life in harsh environments on Earth. Yet this is just what one would expect in the even harsher environment of Mars. Besides, as Benner and colleagues (2000) discuss, even supposing that the GCMS had been functioning properly, it could not have detected organic compounds that might be produced by a modestly different form of carbon-based life. The upshot is that the results of the Viking GCMS are just as ambiguous with regard to the question of Martian life as the results of the LR experiment. In light of these considerations, treating the failure of the Viking GCMS to find organics as settling the interpretation of the anomalous results of the LR experiment is puzzling.[1] Nevertheless, this is what happened. As discussed below, attempts to resolve the combined results of the LR experiment and the GCMS in terms of nonliving geochemical processes has not been entirely successful.

In keeping with the rejection of a biological explanation, NASA has spent the past 40+ years focused on explaining the results of the LR experiment in terms of a

---

[1] Underscoring this point, NASA's Curiosity rover detected complex organic molecules in Martian mudstones (Eigenbrode et al. 2018); whether these compounds are the products of life or abiotic geochemical processes remains an open question.

powerful inorganic oxidant that rapidly converts organic compounds to carbon dioxide; this would account for both the strong initial response of the soil to the LR nutrient solution and the failure of the GCMS to discover organic compounds that must have been delivered to the Martian surface by meteoritic dust and, as we now know, other abiotic processes (Steele et al. 2016). Various candidates have been proposed, including hydrogen peroxide and superoxide (Hunten 1979; Oyama et al. 1977), exotic oxidation states of iron (e.g., Tsapin et al. 2000), and (most recently) perchlorate (Hecht et al. 2009); see Lasne et al. (2016) for a review of this history. The first two candidates failed to live up to their initial promise (ten Kate 2010; Zent and McKay 1994). Perchlorate, which becomes a potent oxidizer above 200 °C, is currently popular. Unexpectedly discovered by the Phoenix Mars Lander in 2008, perchlorate was initially hailed as providing a nonbiological explanation for both the results of the LR experiment and the failure of the Viking GCMS to find organic molecules (Hecht et al. 2009). Because the GCMS heated samples of Martian soil up to 500 °C, perchlorate could explain the failure of the Viking GCMS to find the missing organic molecules. Unfortunately, as chemists quickly pointed out, it could not explain the puzzling results of the LR experiment since the temperature maintained in the test chamber during the experiment was 10 °C, too low for perchlorate salts to become chemically reactive. Efforts to provide a plausible abiotic explanation for the results of the LR experiment in terms of complex geochemical reactions involving perchlorate continue. Quinn et al. (2013) argue that ionizing radiation penetrating the thin atmosphere of Mars could decompose perchlorate salts in the Martian soil into chemically reactive products that could explain the results of the LR experiment. Not everyone agrees, however; see, for example, Lasne et al. (2016) and Levin and Straat (2016). The point is more than forty years after the Viking missions, most scientists remain committed to explaining the puzzling results of the Viking biology investigations in terms of an inorganic oxidant. With a few exceptions (e.g., Blanciardi and colleagues 2012), the possibility that they are the product of a novel biology has not been taken very seriously.

To wrap up, both sides of the controversy over how to interpret the results of the LR experiment fail to acknowledge that they were truly ambiguous with respect to the question of Martian life. The PIs of the LR experiment contended that they found life because the results satisfied the (in essence, defining) criteria for life agreed upon in advance by the Viking biology team. The official NASA position is that it did not find life because some of the results did not conform closely enough to the metabolic behavior of known Earth microbes and, most importantly, the GCMS did not detect any organic compounds. The truth is, even supposing that the results can be explained by perchlorate in the Martian soil, NASA did not at the time know what it found. The results of the Viking investigations were deeply puzzling: They both resembled and failed to resemble familiar Earth life in provocative and unanticipated ways. In short, they represented a potentially (but not definitively) biological anomaly. There are, of course, no guarantees that a potentially biological anomaly is in fact biological any more than there is a guarantee that it is abiological.

8.3 The Role of Anomalies in Scientific Discovery

The history of the controversy over the Viking biology experiments underscores an important point: It is a mistake to treat anomalous results as evidence that a scientific hypothesis under investigation (e.g., there is microbial life on Mars) is false *and* it is also a mistake to ignore them on the grounds that other results technically satisfy a preconceived definition (and hence the hypothesis under investigation is true). For as this section argues, anomalies are the best grist for the theoretical mill, challenging preconceived ideas and paving the way to new and better scientific theories. My account of the role of anomalies in scientific discovery is inspired by Thomas Kuhn's work. There are critical differences, however, between our accounts. Kuhn defends a strong version of the theory ladenness of observation – the view that anomalies cannot even be *perceived* as deviant before the advent of at least a scientific crisis, if not a scientific revolution. Like most philosophers of science, I reject the strong version of the theory ladenness of observation. Indeed, I propose using *tentative* criteria to expedite the recognition of anomalies by empirically undermining the cognitive (versus perceptual) grip of theoretical commitments on the scientific mind.

On Kuhn's account, an anomaly is a violation of a paradigm-induced expectation about a domain of natural phenomena (Kuhn 1970, Ch. 6). Kuhn characterizes a (scientific) paradigm very broadly as encompassing not only theory, methods, instruments, and classic experiments but also widely accepted presuppositions that are not consequences of a universal theory. Paradigms facilitate the construction of hypotheses, design of experiments, and interpretation of results. But they may also hinder the exploration of nature, discouraging certain avenues of research and biasing the ways in which data are interpreted. On Kuhn's account, recognizing that a puzzling result is anomalous, as opposed to merely another puzzle to be solved within the intellectual confines of a reigning paradigm, requires a "crisis," in which the foundations of the paradigm are openly being challenged, or a scientific revolution, in which the paradigm is replaced by a new paradigm. In the absence of either of the latter conditions, Kuhn contends that a dominant paradigm literally blinds researchers to the presence of anomalies. Anomalies are either not perceived as puzzling – they are literally invisible – or alternatively are interpreted as challenging puzzles that will be eventually solved using the resources (concepts, laws, methods, etc.) provided by the paradigm.

Kuhn's best-known illustration of how paradigms blind researchers to the presence of anomalies is William Herschel's discovery of the planet Uranus (Kuhn 1970, 115–117). Between 1690 and 1780 some of the most eminent astronomers in Europe reported seeing a star in positions that we now know were occupied by Uranus. Twelve years later Herschel, using a new, more powerful, telescope of his own devising, observed that the object was disk shaped, the wrong shape for a star. Intrigued, Herschel conducted further observations and discovered that the puzzling object moved among (as opposed to with) the stars. Herschel announced that he had discovered a new comet. Word spread quickly and astronomers from all over Europe trained their telescopes – the same inferior telescopes with which they had previously "seen" a star – on the now puzzling object and concurred that it could not be a star. Efforts to fit its observed motion to a cometary orbit were unsuccessful, however, and within a few months the Scandinavian astronomer Anders Lexell announced that his calculations indicated that the orbit was planetary. As Kuhn discusses, Lexell's announcement was followed rapidly by the discovery of numerous smaller celestial bodies (asteroids) having planetary orbits.

Kuhn contends that astronomers failed to recognize Uranus's nonstar-like characteristics because they were committed to a "minor" paradigm providing them with only two "perceptual categories (star or comet)" for classifying celestial objects beyond Saturn (Kuhn 1970, 116); this paradigm included the assumption that Saturn is the outermost planet in our solar system. In Kuhn's words, "the shift of vision that enabled astronomers to *see* [italics are mine] Uranus, the planet, ... helped to prepare astronomers for the rapid discovery, after 1801, of the numerous minor planets or asteroids" (p. 116).

Kuhn's contention – that the discovery of Uranus illustrates the strong version of the theory ladenness of observation – is problematic, however. If their adherence to only two perceptual categories for classifying celestial bodies *literally* blinded astronomers to perceiving the nonstar-like characteristics of Uranus, why was Herschel able to perceive them using a more powerful telescope? The failure of astronomers to recognize that Uranus has the wrong shape for a star is more readily explained by the low magnifying power of their telescopes, which did not permit them a sufficiently clear image of Uranus to be visually struck by its deviant shape, especially given their cognitive (versus perceptual) belief that there are no planets beyond Saturn. Herschel's more powerful telescope made the anomalous shape of Uranus much more conspicuous. Moreover, noticing that Uranus moves among (versus with) the stars requires extended observations. Given how many stars there are in the night sky, why would astronomers bother to spend much time observing Uranus's motion (*qua* star) unless they already suspected that it was more interesting than other celestial objects within the vast field of stars available to them? Herschel found a reason for looking at Uranus's motion vis-à-vis other celestial bodies when he discerned, through the aid of a more powerful telescope, that the shape of Uranus was anomalous for a star. When Herschel reported his findings to his peers, they found a reason for looking more closely at what had now become an intriguing celestial object and, despite the low magnifying power of their telescopes, were able to discern that he was right about its shape and motion. It is hardly surprising that astronomers subsequently dedicated significant efforts to searching for other planet-like objects. The discovery of Uranus defeated a central tenet guiding astronomical research, namely, that all the planet-like objects had already been discovered – that there are no planets beyond Saturn.

As suggested above, there is an important kernel of truth to Kuhn's account: Widely accepted scientific beliefs make it difficult to *recognize* that what one is observing is anomalous.[2] It is not so much that such beliefs alter our actual perceptions as that our expectations about a phenomenon, coupled with limitations of our instruments, influence how closely we attend to it. Furthermore, despite being credited with the discovery of Uranus, it is Lexell, not Herschel, who recognized that Uranus is a planet. Herschel's contribution was recognizing the anomalous character of Uranus *qua star*. Moving beyond controversial details of Kuhn's account,[3] the story of the discovery of Uranus (the planet) illustrates two critical stages in groundbreaking scientific discoveries: (1) Recognizing that a phenomenon is anomalous, as opposed to merely puzzling, and (2) resolving the anomaly by revising or rejecting one or more widely held scientific beliefs. My focus in this chapter is on the first stage, that is, on expediting the recognition that a phenomenon is anomalous. For as the history of science reveals, such recognition is critical to the development of new and better scientific concepts and principles for understanding a domain of natural phenomena.

Because Kuhn was trained as a physicist, as well as historian-philosopher of science, he emphasized the role of physical theories in suppressing recognition of anomalies. But anomalies are also common in the biological sciences. As the biological cases discussed below illustrate, recognizing the anomalous character of a phenomenon, while challenging, does not require anything as drastic as a Kuhnian crisis, let alone paradigm change.

Thomas Cech's discovery of ribozymes (RNA molecules capable of acting as enzymes), for which he and Sidney Altman shared the Nobel Prize in 1989, provides a salient illustration of how a widespread belief about life that is not

---

[2] This amounts to a weak (nonperceptual) version of the theory ladenness of observation, which is the version most widely accepted by contemporary philosophers of science.

[3] Kuhn's account of scientific revolutions has been heavily criticized for, among other things, the vagueness of the concept of paradigm, which infects the viability of his account of scientific revolutions: What counts as a "minor" paradigm? What counts as a genuine paradigm shift (scientific revolution)? It is beyond the scope of this chapter to delve into the details of these problems, especially since they do not affect the account that I am defending of the role of recognition of anomalies in facilitating trailblazing scientific discoveries.

grounded in a general theory can delay an important scientific discovery. The discovery of ribozymes challenged what Francis Crick (1958) dubbed the "central dogma of molecular biology." According to the central dogma, genetic information flows in one direction from DNA to RNA to protein and, most importantly for purposes of this discussion, there is a rigid division of labor between nucleic acids (DNA and RNA), which store, transfer, and manage genetic information, and proteins, which catalyze biochemical reactions. Cech and his team were not looking for ribozymes. They fortuitously stumbled upon RNA molecules that were catalyzing reactions like protein enzymes. The recognition that this was happening took time, however, because it violated the central dogma.

Cech's team was investigating a newly discovered phenomenon known as RNA cutting and splicing, which occurs when an RNA molecule is being assembled from a DNA template. The process involves cutting out and tossing away pieces of the precursor RNA molecule and then splicing the ends back together to form a biologically functional RNA molecule. Little was known about how splicing takes place. In keeping with the central dogma, Cech assumed that a protein catalyst (enzyme) was responsible. His goal was to isolate it. The basic idea was to add enzymes to a purified sample of precursor RNA until they found one that spliced it. Cech's team acquired (unspliced) precursor rRNA by synthesizing it from genes (rDNA) for rRNA acquired from a ciliated protozoan, *Tetrahymena*; see Cech (1989) for more detail. To their surprise, splicing occurred in a preparation that they believed lacked protein catalysts. Extensive efforts to remove proteins from precursor rRNA samples failed to halt the splicing. They were facing an anomaly for which there was no obvious explanation. In desperation they conjectured that the enzyme must be very tightly bound to the RNA, which seemed implausible given the chemical and physical "abuse" to which they had subjected the sample in their efforts to remove proteins. As Cech notes, "[t]hat we took this hypothesis seriously provides an indication of how deeply we were steeped in the prevailing wisdom that only proteins were capable of highly efficient and specific biological catalysis" (Cech 1989, 659). At a loss, Cech and his students put the project of isolating the splicing enzyme aside for a time and focused on other aspects of the splicing reaction. Gradually, over the course of almost a year and a half, Cech and his students reached the conclusion that the RNA molecule itself must be the splicing agent (Kruger et al. 1982).

The grip of the central dogma of molecular biology on the minds of Cech and his students delayed the discovery of ribozymes. *In hindsight*, the time to have considered this possibility was when even extreme measures failed to stop the splicing. Instead they entertained an implausible hypothesis about a protein catalyst being so tightly bonded – as Cech notes, "perhaps even covalently bonded" (Cech 1989, 659) – to the precursor rRNA molecule that even extreme measures, such as boiling in detergents at high temperatures and treatment with several nonspecific proteases (which break peptide bonds), failed to stop the splicing. The similarities between this case and the Viking biology investigations are striking. Just as Cech's team sought to explain the anomalous splicing of precursor RNA in terms of an increasingly mysterious protein catalyst, so most astrobiologists attempt to explain the puzzling results of the LR experiment in terms of a mysterious inorganic oxidant. What the astrobiology community has not done, and Cech's team finally did, is take seriously the possibility that the key presupposition undergirding their reasoning is mistaken.

This is not to claim that the anomalous results of the LR experiment were produced by Martian microbes; perchlorate in the Martian soil is a promising candidate for explaining them. It is merely to point out that persistent failure of extensive efforts to explain a puzzling finding (discovered in the laboratory or through field research) within a framework of widely accepted scientific beliefs renders the phenomenon under investigation anomalous. And this, in turn, suggests that something may be wrong with the set of beliefs concerned. Although not widely appreciated, Carl Sagan's oft-cited quote, "extraordinary claims require extraordinary evidence," applies not only to the hypothesis that the results of the LR experiment are due to an unfamiliar form of life but also to the conjecture that they were produced by a mysterious inorganic oxidant; until the anomaly is truly resolved, the latter conjecture is no less extraordinary than the former.

In keeping with the received interpretation of the results of the Viking biology experiments, the focus of subsequent life detection missions to Mars has been on searching for signs of ancient Martian microbes – microbes that may have lived several billion years ago, during a time when Mars is thought to have been warmer and wetter than it is today; the notable exception is NASA's ill-fated Mars 96 mission, which included an instrument package, MOx, for investigating whether the oxidant responsible for the Viking results could be biological.[4] A two-part life detection mission to Mars, dubbed ExoMars (Exobiology on Mars), jointly funded by the European Space Agency (ESA) and Russian Federal Space Agency (Roscosmos), is in progress. The first half of the mission (ExoMars Trace Gas Orbiter) was launched in the spring of 2016,[5] and the second half (ExoMars Rover), which will search for chemical and morphological signs of past microbial life, will be launched in 2020. NASA too is funding a rover mission, known as Mars 2020, to search for signs of ancient Martian microbes. In addition to an

---

[4] Additional missions to investigate the possibility of extant life on Mars have been proposed but not funded. A good illustration is BOLD (Biological Oxidant and Life Detection). The stated objective of BOLD is to ". . . directly address the presence of the 'mysterious' oxidizer on the Martian surface and any biological activity" (Schulze-Makuch et al. 2012).

[5] Unfortunately, the Schiaparelli EDM lander crashed but the Trace Gas Orbiter is in orbit around Mars.

instrument package (SHERLOC) for searching for signs of past microbial life, the Mars 2020 rover is equipped with a drill for acquiring subsurface samples of Martian rock and soil that will be collected and preserved for return to Earth in a future mission; returning samples to Earth will permit a much more thorough investigation for signs of ancient Martian life than can be done in situ on Mars.

There are many other cases in the biological sciences where commitment to widely accepted beliefs delayed recognition that a puzzling empirical finding is anomalous. The discovery of the Archaea, discussed in Section 5.3.2, provides a good illustration. By the middle of the twentieth century, the prokaryote–eukaryote distinction was widely accepted by biologists as evolutionarily significant. Taxonomists classified all prokaryotic organisms, which are unicellular and lack a chromosome containing cell nucleus, together in a single Kingdom "Monera" (bacteria) and all "true" (excluding fungi) unicellular eukaryotes, which have a chromosome containing cell nucleus, into a separate Kingdom "Protista." The assumption that the prokaryote–eukaryote distinction is phylogenetically significant was based upon the seemingly plausible supposition that one can infer genetic relationships from gross morphological characteristics, in the case of unicellular microbes the distinctive internal structure of their cells.

In the latter part of the twentieth century, with the development of sophisticated methods for analyzing genomic sequences, this supposition began to unravel. Woese and Fox (1977) collected a large database of ribosomal RNA sequences from a wide variety of prokaryotes. To their surprise some of these "bacteria" differed genetically from each other in their transcription and translation machinery more than they differed from eukaryotes. Moreover, the lipids in their cell membranes were also anomalous; as discussed in Section 5.3.2, they differed not only from other bacteria but also from eukaryotes. Efforts to accommodate these puzzling discoveries within the Five Kingdom taxonomic scheme for life proved difficult. In 1990, Woese took a radical step and boldly proposed restructuring the highest level of biological systematics into three "Domains" of life (Archaea, Bacteria, and Eukarya) (Woese et al. 1990). What is important for our purposes, however, is that it is clear, in hindsight, that there were telling signs that prokaryotes do not form a natural phylogenetic category. But because microbiologists were working under the prokaryote–eukaryote evolutionary paradigm these signs went unrecognized for what they represent. As students were counseled in a late 1970s textbook on thermophilic (heat loving) bacteria:

The fact that *Sulfolobus* and *Thermoplasma* have similar lipids [in their cell membranes] is of interest, but almost certainly this can be explained by convergent evolution. This hypothesis is strengthened by the fact that *Halobacterium*, another quite different organism, also has lipids similar to the acidophilic thermophiles.

(Brock 1978, 178)

All three of these "bacteria" are archaea.

Analogous to the discovery of numerous small planetoids following the discovery of Uranus (*qua* planet), the discovery of the archaea was followed by the discovery of increasingly divergent genomic sequences in environmental samples as more sophisticated metagenomics tools were developed for exploring the microbial world. Some of the sequences discovered in recent years are so divergent from known microorganisms that there is speculation that they may represent as yet undiscovered (fourth, etc.) domains of Earth life (Wu et al. 2011). Whether these mysterious genomic sequences are derived from unknown cellular microbes or viruses is unclear. Most microbiologists attribute them to unknown viruses but are nonetheless divided as to whether these (hypothesized) viruses evolved as intracellular parasites from a now extinct fourth (etc.) domain of cellular life or (given the discovery of complex, bacteria-like, giant viruses, discussed below) should themselves be classified as noncellular forms of life (Moreira and López-García 2015). The point is the discovery of the archaea opened up new conceptual possibilities for thinking about the evolutionary structure of the microbial world.

My last illustration of how widely shared assumptions can blind biologists to the significance of puzzling empirical findings is the discovery of giant viruses. The first giant viruses (Mimivirus) were found in amoebae isolated from a water-cooling tower during an outbreak of Legionnaires' disease in 1992; see Wessner et al. (2010) for the history. Biologists did not, however, identify them as viruses. They violated a widespread belief, based on observation and theoretical considerations, about how large a virus could be. At the time, all known viruses were much smaller than bacteria. Their small size was explained theoretically in terms of their greater functional simplicity. Viruses reproduce by commandeering the genetic and metabolic machinery of a host cell, and hence do not need the complex molecular machinery (ribosomes, cell membranes, cell walls, etc.) of bacteria, which are self-reproducing. They consist of little more than protein coated DNA or RNA molecules. The "particles" discovered inside amoebae were huge compared to known viruses, with a total diameter of approximately 750 nm (including fibers extending from their capsids), larger than some parasitic bacteria. In addition, they stained gram-positive, reinforcing the notion that they were bacteria (Birtles et al. 1997).[6] Mimiviruses were thus initially classified as a new variety of small bacteria. It was not until ten years later that they were recognized as giant viruses (La Scola et al. 2003). Reminiscent of the discovery of many small planetoids after the discovery that Uranus is a planet, several larger

---

[6] Gram staining is used to divide bacteria into two large groups, gram-positive and gram-negative, on the basis of the chemical properties of their cell walls. It is typically the first step in the process of classifying a bacterium.

viruses – pandoraviruses (Philippe et al. 2013) and megaviruses (Arslan et al. 2011) – have since been found.

There are echoes of the initial interpretation of mimiviruses as bacteria in the controversy over the interpretation of mysterious structures discovered in the famous Allan Hills (Martian) meteorite. Found in Allan Hills, Antarctica in 1984 by meteorite hunters, ALH84001 – as it was dubbed by virtue of being the first meteorite collected during that expedition – was identified in 1993 as Martian on the basis of trapped traces of stable atmospheric gases; the composition of the Martian atmosphere, which is significantly different from Earth's, is known from the Viking missions. In 1996 David McKay and colleagues (McKay et al. 1996) decided to investigate the compositions of the few (just twelve) Martian meteorites that had thus far been found. They discovered carbonate mineral globules in ALH84001 dating to around 3.9 bya, making them much older than the impact event that blasted ALH84001 from the surface of Mars and sent it on its way to Earth. Carbonate globules form in aqueous environments. The presence of carbonate globules in ALH84001 was consistent with what was known about the early climate of Mars. Until around 3.8 bya, Mars is thought to have been wetter and warmer, with a much thicker atmosphere. What most excited McKay and his team, however, was what they found inside the carbonate globules. A scanning electron microscope revealed small sausage-shaped structures closely resembling fossilized Earth bacteria. Even more provocatively, the carbonate globules contained miniature magnetite crystals whose size, chemical purity, and uniform geometric shape closely resemble the tiny magnetite crystals produced by magnetotactic bacteria on Earth (McKay et al. 1996; Thomas-Keprta et al. 2000). Magnetotactic bacteria produce chains of tiny magnetite crystals and use them, like tiny compasses, to orient along Earth's magnetic field in boggy environments (Faivre and Schüler 2008). When found in ancient Earth rocks, magnetite crystals having these characteristics are interpreted as microfossils (remnants of long dead, magnetotactic bacteria) (Kopp and Kirschvink 2008). The magnetite crystals provide the most suggestive evidence that the structures in the carbonate globules are the remains of ancient Martian life.

On the other hand, there are puzzling differences between the structures found in ALH84001 and those associated with familiar Earth microorganisms. One of the more influential arguments against the conjecture that the sausage-shaped structures found in the carbonate globules are fossilized Martian microbes is their very small size, around 20–100 nm in diameter. In contrast, the smallest known Earth bacteria are around 200 nm in diameter. Many biologists (e.g., Knoll et al. 1999) argued that they are too small to contain the genetic and metabolic machinery required by life, a claim that clearly rests upon observations of known cellular life on Earth. Put provocatively, just as mimiviruses were initially rejected as too big to be viruses, so the bacteria-shaped structures found in ALH84001 are rejected as too small to be fossilized Martian microbes.

8.4 Searching for Anomalies Using Tentative (Versus Defining) Criteria

Most puzzling scientific findings do not give rise to major discoveries such as ribozymes, archaea, and giant viruses. Ingenuity and hard work eventually resolve them within the context of widely accepted scientific beliefs. On the other hand, as the cases just discussed underscore, major scientific discoveries often begin with findings that resist efforts to explain them within a prevailing theoretical framework. At what stage do unanticipated findings cross the threshold from being merely puzzling to anomalous? This issue is especially important for those searching for extraterrestrial life because, as the Viking biology experiments and studies of ALH84001 illustrate, not much data are available and access to pertinent phenomena is quite limited. Astrobiologists want to put their money (literally) on the most promising candidates for extraterrestrial life. Following earlier work (Cleland 2012, 2019; Cleland and Chyba 2007), this chapter recommends using tentative (as opposed to defining) criteria to search for extraterrestrial life.

Unlike defining criteria, the purpose of tentative criteria is not to settle the question of whether an extraterrestrial environment harbors life but instead to evaluate whether it contains phenomena that are anomalous in a biologically promising way – phenomena that resist efforts to classify them as biological or abiological. Universal characteristics of familiar life that are widely viewed as fundamental to life should be included as tentative criteria in a search for extraterrestrial life. Because they are interpreted as provisional, however, the thorny question of which is fundamental to life will not arise. Instead of competing, as is typical with popular metabolic and evolutionary definitions of life (Section 1.5.1), allegedly "essential" characteristics of life may function jointly in a set of criteria for searching for extraterrestrial life. The proposed strategy diminishes the risk of prematurely labeling weird life as *ipso facto* nonbiological on the grounds that it fails to exhibit a characteristic mistakenly thought to be fundamental to life. If we are wrong about a characteristic being basic to life – if unbeknown to us it is contingent upon conditions present during the origin and early evolution of life on Earth – other criteria may still flag the phenomenon as suspiciously biological.

It is important that all the criteria used in a search for biologically promising anomalies be regarded as defeasible, that is, as revisable and even eliminable in light of new scientific discoveries and theoretical developments. Our current experience with life is limited to a single example, and as Section 5.2 discussed, the worry that familiar life may not be representative is not merely logical; there are scientific reasons, both empirical and theoretical, for concern as well. Astrobiologists are faced with the very real (versus merely logical) possibility that the metabolic-based and genetic-based strategies used by familiar life are symptoms of (as yet unrecognized) more fundamental, underlying processes. Such life conferring processes might play out differently under chemical and physical conditions different from those on Earth. As an analogy, tuberculosis was once identified on the basis of symptoms produced by an unknown, underlying infectious agent, the bacterium MTB (*Mycobacterium tuberculosis*). MTB infects both the lungs and the lymph system but produces quite different symptoms, respectively, a blood-stained cough versus tumor-like swellings on the neck. On the basis of symptoms alone, tuberculosis was identified as two different diseases – consumption, a lung disease, and scrofula, a disease of the lymph nodes – before the advent of the germ theory of disease. Just as the symptoms of an MTB infection vary, depending upon which organ of the body is infected, so any of the putatively universal characteristics of known life may reflect contingencies present at the time of the origin of life or during its early evolution. It is thus important that all (including allegedly fundamental) characteristics of life be treated as revisable in light of future empirical discoveries and theoretical developments. Otherwise astrobiologists run the risk of failing to recognize truly novel life should they be fortunate enough to encounter it.

It is also important that tentative criteria be deployed jointly in a search for potentially biological anomalies. The set of tentative criteria used for screening an extraterrestrial environment for novel life should include contingent as well as universal characteristics of familiar life. An extraterrestrial system satisfying enough provisional criteria for life will qualify as worthy of further investigation for life even supposing that it fails to satisfy an allegedly fundamental characteristic of life; the latter is merely one, perhaps heavily weighted but certainly not conclusive, characteristic to be taken into consideration in assessing a geochemical system's potential for life. What constitutes "enough" will vary depending upon the individual case and, most importantly, the *beliefs* of the scientists involved. For whether a phenomenon is viewed as anomalous depends upon the current state of our scientific knowledge; once a phenomenon is understood it ceases to be puzzling.

A challenging, often underappreciated, problem facing astrobiologists is that of *operationalizing* characteristics of familiar life as criteria (tests) for life. As discussed in Section 4.3, the most fruitful scientific theories maximize scope (generality) while minimizing exceptions, and this requires hitting the "right" (Goldilocks) level of abstraction. The same is true for operationalizing characteristics of familiar life for the purpose of searching for novel life. One wants to avoid either end of the abstract–concrete continuum. Tailor a search criterion too closely on characteristics of familiar life and you will not be able to recognize weird forms of life. Abstract too far away from the biological details of familiar life and many nonliving systems will qualify as promising candidates for life. As an illustration of the latter worry, one of the most striking characteristics of familiar life is metabolism, that is, the capacity of living things to sustain themselves in a far from equilibrium state by extracting energy from their surroundings. Metabolism can be rendered so abstract as to qualify purely informational computer simulations as "metabolizing"; the same is true for genetic-based Darwinian evolution, which as discussed in Chapter 7, is even easier to treat as purely informational than metabolism. Metabolic definitions of life based on chemical thermodynamic considerations attempt to capture a less abstract notion of metabolism but face the problem that some nonliving material systems maintain themselves in a state of disequilibrium by extracting energy from their environments. Tornadoes, Bénard cells, and the great red spot of Jupiter provide salient illustrations. A highly effective, thermodynamic based criterion for searching for novel life would be even less abstract, able to distinguish biological from abiological states of thermodynamic disequilibrium. Coming up with such a criterion is not an easy task, however, given our limited experience with extraterrestrial phenomena. Until fairly recently it was widely believed that the only material systems that can be far out of thermochemical redox equilibrium were biological. As Seager and Bains (2015) discuss, however, recent studies indicate that there are plausible planetary conditions in which abiotic processes produce such disequilibria; I will have more to say about this shortly.

On the other hand, criteria that are too closely based on familiar Earth life run the risk of being too specific to identify natural variations in the biological characteristics that they are designed to detect. The Viking mission LR experiment used the ability to metabolize a specific nutrient solution as a metabolic criterion for life. The nutrient solution was selected because it was thought to be metabolizable by "most" bacteria (those that had been cultivated in the laboratory). Since the time of the Viking missions, however, biologists have discovered (using cultivation-independent molecular techniques) that the vast majority (99%) of bacteria and archaea cannot be grown in laboratory culture (Hugenholtz et al. 1998) and moreover that many of these microorganisms (especially archaea) "feed" exclusively on inorganic compounds (hydrogen gas, ferrous iron, etc.). The upshot is that we now know that most Earth microorganisms could not be cultivated under the conditions provided by the LR experiment. In retrospect, the use of the LR nutrient solution in a metabolic criterion for life seems very naïve; even at the time there was little reason to think that Martian microbes would be similar enough to bacteria in their biochemical pathways to metabolize it. As Schuerger and Clark (2008) discuss, future life detection missions to Mars are likely to use a variety of nutrients, tailored to what we currently know about the metabolic strategies of microbes living in Mars analog environments.

In sum, the potential fruitfulness of a search criterion for novel life depends upon avoiding extremes of abstractness and specificity. Failure to satisfy a criterion might not be evidence that the associated biological characteristic (e.g., metabolism) is absent but instead evidence that the criterion (e.g., ability to metabolize the LR nutrient solution) is too specific to known Earth life to provide a good test for that characteristic. Likewise, satisfaction of a criterion might reflect a problematic translation of a biological characteristic, such as metabolism, into a criterion (e.g., thermochemical redox equilibrium) that is too general to exclude abiotic phenomena in certain environmental contexts. Given our current epistemic limitations, the best strategy is to avoid criteria at either extreme of the abstract–concrete continuum, and to view all criteria as truly tentative – to be open to revising them (either upward or downward along the continuum) in light of new empirical discoveries and theoretical developments. In this context, it is worth noting that advocates of a definitional approach are especially prone to conflating (allegedly "essential") characteristics of life with the search criteria designed to detect them. Levin's continued insistence that the Viking missions discovered Martian life, on the grounds that the agreed upon pre-mission criteria for metabolism were technically satisfied, provides an especially salient illustration. We now know that the metabolic search criteria used in the LR experiment were too specific to encompass the diversity of types of metabolism exploited by familiar Earth microbes, let alone extraterrestrial microbes.

So what characteristics of familiar Earth life might be successfully operationalized as tentative criteria in a search for biologically promising anomalies? Once one recovers from the perceived need to "define" life, many candidates come to mind. Any characteristic universally found in association with familiar life and, most importantly, rarely (if ever) found in association with nonliving phenomena is a promising candidate. Highly contingent, nonuniversal characteristics associated with specialized biological adaptations to particular types of environments also make promising candidates. For they too stand out against the backdrop of nonliving geochemical processes, and in addition hold forth the promise of flagging chemically different forms of life facing similar functional (adaptive) challenges.

Somewhat surprisingly, given the lip service commonly given to "defining" life, many such characteristics are already either in use or being considered for use in life detection missions. They span the range from structural, to compositional, to functional. A brief survey of some of them is revealing.

Many microorganisms found on Earth live in microbial communities, such as biofilms (Stubbendieck et al. 2016). It has thus been suggested (e.g., Cady and Noffke 2009; Vago et al. 2017; Westall 2008) that if Mars hosted life around

3.9 bya, when it was much warmer and wetter, it may have produced microbially induced, communitarian, sedimentary structures, such as stromatolites, and that fossilized traces of them might still exist. Such structures have the advantage of being easier to detect than the fossilized remains of solitary microbes. There are no guarantees, however, that microbes found in extraterrestrial environments somewhat (superficially?) analogous to those found on Earth will be as communal as those found on Earth; this is the old philosophical conundrum of evaluating which respects of similarity and difference are causally relevant in a particular case. But even supposing, for the sake of argument, that all microbes are as gregarious as many of those found on Earth, it is unclear what sorts of structural traces their communities might leave in an extraterrestrial environment; they might be unrecognizable as products of biology. Nevertheless, sedimentary structures of the sort produced by familiar microbes make promising tentative criteria for searching for extraterrestrial life on a planet like Mars. The point is that they should be treated as tentative and the results evaluated in the context of a suite of other criteria.

Proteins and nucleic acids are notable for their extremely large size, polymeric architecture, and complexity compared to molecules produced by known nonliving geochemical processes. Indeed, before the discovery of DNA, Schrödinger predicted, on theoretical grounds, that any molecule capable of storing the hereditary information required for Darwinian evolution would be a large aperiodic crystal; such molecules are not found in nonliving geochemical systems (Schrödinger 1944, 5). The discovery of extremely large and complex polymers, regardless of their chemical composition, would thus be suggestive of life. The biomolecules of familiar life also exhibit distinctive patterns of structure and composition that could prove useful in a search for novel life; Chris McKay (2004) dubs this the "Lego Principle." They are built from small subsets of classes of chemically similar organic molecules. As an illustration, the proteins of familiar life are synthesized from 20 L-amino acids (that are directly genetically encoded) even though natural processes produce over 100 amino acids of mixed (L and D) chirality. Thus, a mysterious enrichment in a small subset of homochiral amino acids, regardless of whether they are the same as those used by familiar life, is suggestive of biology. On the other hand, there may be as yet poorly understood abiotic processes that produce similar enrichments in both organic and inorganic molecules, especially under conditions very different from those found on Earth. The point is while McKay's Lego Principle provides a promising tentative criterion for life it would be a mistake to hold it up as a definitive criterion.

Some biologists, for example, Pace (2001), go even further, contending that there is one best way of doing biochemistry and that natural selection will always find it, and hence that alien biochemistries will resemble ours in chemical composition as well as structure. Given our limited experience with a single example of life, however, such a highly specific theoretical speculation about life elsewhere in the universe seems very premature. Underscoring this point, McKay (2016) speculates that if there is life on Titan it will be carbon based but (contra Pace) biochemically different from familiar life, and Schulze-Makuch and Irwin (2008) and Bains (2004) conjecture that (contra McKay) silicon might be capable of forming a sufficiently diverse set of complex polymers for life under conditions such as those found on Titan, namely, extremely low temperatures and lacking both liquid water and oxygen. Who knows which of these theoretical speculations about the possibilities for life elsewhere in the universe is more plausible? Instead of designing criteria founded upon detailed but tenuous speculations about what life *might* be like in an environment as strange as Titan's, a more promising strategy is simply to screen for anomalies, such as large and complex aperiodic polymers or, following McKay's Lego Principle, mysterious patterns of concentrations of chemicals. The latter strategy holds forth the promise of flagging (for additional scientific investigation) novel biochemistries that have yet to be dreamt up by Earth-bound biochemists, and if carbon or silicon biochemistries are present on Titan it should flag signs of them as well.

Familiar life preferentially builds its biomolecules from the lighter isotopes of certain elements, most notably, carbon, nitrogen, and sulfur. This is not surprising when one considers that lighter isotopes increase the efficiency of biochemical processes, and hence are likely to be strongly selected for by evolutionary mechanisms. The biomolecules of any form of life are thus likely to exhibit a preference for the lighter isotopes of elements. There are environmental conditions, however, in which abiotic processes produce enrichments in the lighter isotope of an element. As an illustration, the lighter isotope of oxygen, ^{16}O, evaporates more rapidly than the heavier isotope, ^{18}O, and hence is a good signature for climate change on a world, such as Earth, in which all three phases of water are present. The point is the physicochemical nature of an environment plays a significant role in determining whether an enrichment of the lighter isotope of an element is biologically suspicious. This underscores the importance of tailoring criteria for searching for novel life to particular environmental contexts. As Seager et al. (2016, 466) observe, "habitability is planet-specific."

Certain patterns of distribution and concentration of rare minerals are also associated with biological processes, and hence make promising candidates for signaling potentially biological anomalies. Of the more than 5300 mineral species known on Earth, fewer than 100 comprise 99% of Earth's crustal volume. Most are rare (occurring at five or fewer localities) and more than half of them are biogenic (produced or mediated by biological processes), suggesting an intimate association between life and rare mineral species (Hazen et al. 2008; Hystad et al. 2015).

The banded iron formations, which contain large deposits of hematite, an iron oxide mineral marking the advent of the oxygenation of Earth's atmosphere around 2.45 billion years ago by photosynthetic microorganisms (Canfield 2005), provide a familiar illustration; dioxygen (O_2) is a reactive gas that would rapidly disappear from the atmosphere if it were not constantly replenished by photosynthetic organisms. Thus, mysterious deposits of rare mineral species, that are difficult to explain in terms of familiar geochemical processes, provide promising criteria for past life, suggesting that reactive biogenic gases, such as oxygen or methane (but others should be considered as well), may once have been present in the atmosphere–ocean system of a planet or moon. Other examples of minerals that can form biogenically include the uranium ore uraninite, travertine and opal, which are often deposited through algal activity, the sulfides pyrite and marcasite, which are commonly produced by sulfate reducing bacteria, and many sulfate, carbonate, and phosphate mineral species. While not definitive of life, the presence of unusual concentrations of chemically related, rare minerals in an extraterrestrial environment where they would not be expected, given our current understanding of prevailing geological conditions, is certainly suggestive of biology, and may even hint at the possibility of truly novel metabolisms and community structures.

Highly contingent (nonuniversal) characteristics of known life that are rarely found in nonliving natural systems on Earth also make promising tentative criteria for life. The chemically pure, single domain magnetite crystals found in meteorite ALH84001 provide an especially promising candidate since, as discussed in Section 8.3, similar magnetite crystals are routinely interpreted by paleomicrobiologists as microfossils. Nevertheless, most astrobiologists currently reject the hypothesis that they are biogenic. Magnetite crystals of similar chemical purity, size, and shape have been produced abiotically under conditions deemed analogous to those undergone by ALH84001 (Bell 2007; Golden et al. 2001), and there is reason to believe that the associated organic carbon compounds have an abiotic origin (Steele et al. 2016). This does not, however, alter the point that characteristics unique to a variety of Earth life found in unusual environments provide promising candidates for tentative criteria for searching for life in analogous extraterrestrial environments (until there is evidence that they do not).

The examples canvassed above provide a snapshot of the diversity of characteristics of familiar life that have been suggested (in varying degrees of abstractness and specificity) as criteria for searching for life on planets and moons in our solar system. Searching for evidence of biology on exoplanets orbiting nearby stars is even more challenging since researchers are pretty much limited to analyzing atmospheric gases. Only a few of the thousands of different gases produced by Earth life are detectable remotely by space telescopes, namely, dioxygen (O_2), methane (CH_4), and nitrous oxide (N_2O). This raises the question of which gases astrobiologists should look for in the atmospheres of exoplanets. Seager and colleagues (2016) argue that there are ecological and environmental contexts in which biogenic gases that are rare in Earth's atmosphere might be present at high enough levels to be remotely detected, and hence that it would be a mistake to restrict a search for life to biogenic gases prominent in Earth's atmosphere. Even more intriguingly, they generate a list of "potentially biogenic" small molecules not produced by Earth life that might accumulate in large enough quantities in the atmospheres of exoplanets to be detectable by space telescopes. They admit that their list is excessively general, including all small molecules (organic and inorganic) that are stable and potentially volatile, and argue that it should be further refined in light of hypothetical planetary atmospheres and surface environments in order to be useful for screening exoplanetary atmospheres remotely for the possibility of novel life. Given how little we know about the possibilities for life in environments quite different from those found on Earth, a better strategy might be to screen an exoplanet's atmosphere for anomalous accumulations of potentially biogenic gases, as opposed to trying to anticipate in advance which potentially biogenic gases should be present in an exoplanet's atmosphere if it harbors novel life.

As mentioned earlier, the strategy for searching for life proposed in this chapter resembles the conventional approach of NASA and ESA insofar as it recommends using sets of diverse criteria tailored to particular environmental contexts. An important difference – in addition to the tendency to (at least implicitly) treat criteria for life as definitive, as opposed to provisional – is the focus of the conventional strategy on identifying life *per se*, as oppose to biologically promising anomalies. As Cady and Noffke (2009) observe, the central challenge in "... extraterrestrial life detection is determining whether the phenomena (or suite of phenomena) can be uniquely attributed to life" (p. 4). Similarly, in the context of a discussion of the ExoMars mission, Vago et al. (2017) ask whether "... there is a pragmatic set of robust measurement that could provide proof of life? Better yet, can we devise a scale or scoring system to help us quantify how confident (or otherwise) we have a right to be?" (p. 42). These astrobiologists are not hunting for biologically promising anomalies worthy of more in-depth investigation for novel life. They are seeking sets of criteria that come as close as possible to giving them a definition-like (thumbs-up or thumbs-down) answer to the question of whether an extraterrestrial environment harbors life. Such an approach amounts to looking under the proverbial streetlight for extraterrestrial life: They will find it only if it closely resembles familiar life in ways being privileged by their search criteria. The most interesting forms of extraterrestrial life – those that challenge our current Earth-centric concepts – will be missed because they will fail to satisfy some of these criteria in baffling ways, and hence be relegated to poorly understood abiological processes.

It would be nice if, along the lines suggested by Vago et al., a set of tentative criteria, $C = \{c_1, c_2, \ldots, c_n\}$, could be deployed in a decision procedure (algorithm) for estimating the degree of anomalousness of an extraterrestrial geochemical system vis-à-vis the question of life. The very nature of anomalous phenomena makes this difficult, however. Anomalies violate widely accepted scientific belief in *unanticipated* ways, which makes it difficult to design a probability measure, such as a Bayesian algorithm, for screening for them *in advance* of encountering them. The puzzling result of the LR experiment provides a good illustration. Who could have anticipated, in advance of its occurring, that 30% of the $^{14}CO_2$ gas released after the first injection of the LR nutrient solution would disappear after a second injection? The point is the more anomalous (baffling) a finding the less likely that a mechanical decision procedure will classify it as probable enough to be worthy of more in-depth scientific investigation for life. It is thus a mistake to think that one can search for biologically promising anomalies using an algorithm. Recognizing which anomalies are worth pursuing as biologically promising requires considered human judgement.

The criteria in a set of tentative criteria designed to probe an extraterrestrial system for potentially biological anomalies are not exhaustive. They may be inadequate for the task due to unrecognized limitations of our scientific understanding of life. No one on the LR team anticipated that exposure to a second helping of the LR nutrient solution would result in reabsorption of significant quantities of previously released CO_2 gas; it was too farfetched a possibility for them. As a consequence, they did not include criteria to discriminate biological from abiotic oxidants. Furthermore, the tacit assumption that criteria in a set of tentative criteria C are mutually compatible (can be satisfied together) is highly problematic. We cannot determine in advance that criteria for life that are compatible on Earth are also compatible in extraterrestrial environments. Some of the universal characteristics of Earth life that we currently view as fundamental may turn out to be contingent on geochemical conditions at the time of the origin or early evolution of life on Earth. Would classic Darwinian evolution by natural selection, which as discussed in Chapter 6, is especially useful for adapting to environmental changes that are rapid but not catastrophic, evolve in an extraterrestrial environment that is far more stable than that of Earth? What about an environment that is far less stable than Earth's? The point is the extent of our ignorance about the nature of life, and how it arises and evolves in environmental conditions different from those found on Earth, makes it difficult to design a well-defined probability measure – a measure satisfying the axioms of the mathematical theory of probability – for identifying biologically promising anomalies.

As the discoveries of ribozymes, archaea, and giant viruses underscore, recognizing that a puzzling phenomenon is mysterious enough to challenge widely held scientific tenets requires human judgement. The advantage of the proposed informal strategy for searching for extraterrestrial life using tentative criteria is that it directly confronts scientists with the anomalousness of an extraterrestrial phenomenon vis-à-vis their Earth-centric beliefs about life. They are then faced with the task of explicitly deciding whether the phenomenon is provocative enough to warrant a more thorough investigation into the possibility of alien life, as opposed to rejecting the possibility of life out of hand because it fails to satisfy a privileged definition of life. Such an investigation may not result in the discovery of novel life. Researchers may discover that the tentative criteria concerned are satisfied by abiotic systems under physicochemical conditions not found on Earth. While disappointing, this would nonetheless represent genuine scientific progress in our understanding of life. Astrobiologists would have learned that certain features of life, previously thought to be either universal or at least rare among nonliving systems, are not reliable indicators of life after all.

8.5 Concluding Thoughts

Anticipating how extraterrestrial life might differ from familiar Earth life is one of the most challenging problems facing astrobiologists. Biologists have discovered that life as we know it on Earth represents a single example, and as Section 5.2 discussed, there are reasons for worrying that it may be unrepresentative of life considered generally. This problem is exacerbated when one considers that the most common form of life in the universe is almost certainly microbial; even on Earth, large multicellular organisms are rare and exotic latecomers, in essence, biological outliers. For all we know, alien microbes might resemble familiar microbes closely or they might differ from them in ways that, given our limited, Earth-centric, understanding of life, we cannot anticipate. This makes it difficult to design a strategy for searching for extraterrestrial life using biosignatures based on familiar life.

Despite their popularity, definitional approaches are especially prone to this difficulty. While useful for finding life as we know it, defining criteria are poorly suited for discovering the most theoretically interesting forms of alien life, namely, life differing from our own in surprising and unanticipated ways. Indeed, as discussed earlier, the use of defining criteria is more likely to disguise the presence of novel life than to flag it for further scientific investigation. The challenge is moving between the horns of this dilemma – figuring out how to minimize the chances of overlooking truly novel life while using search criteria based on the only kind of life with which we are familiar. The solution, advocated in this chapter, is to use sets of tentative (versus defining) criteria to identify potentially biological anomalies – phenomena that resemble familiar Earth life in provocative

ways and yet also differ from it in perplexing ways – for further scientific exploration. For as the history of science reveals, anomalies provide the best grist for the
theoretical mill, challenging preconceived ideas, and paving the way to important
scientific discoveries and new and better scientific theories.

Unlike defining criteria, tentative criteria include characteristics of familiar life,
both universal and contingent (upon special conditions), that are not thought to be
fundamental to life. The objective is not to settle the question of whether an
extraterrestrial environment harbors life, although this could happen if it contains
life closely resembling Earth life; this highlights the fact that the strategy being
proposed is *not inferior to a definitional approach but instead improves upon it.*
The goal is to flag a geochemical system as scientifically worthy of a more
comprehensive investigation for life. This will undoubtedly disappoint those
longing for decisive answers to the question of whether an extraterrestrial environment harbors life. But the expectation that we could immediately identify novel life
using criteria based on familiar Earth life is naïve. Most importantly, however, it is
those phenomena that are most difficult to classify as living versus nonliving that
hold forth the greatest promise of moving us beyond our Earth-centric concepts of
life, paving the way to a better understanding of the nature of biological systems.
For as discussed in Chapter 4, the key to formulating a universal theory of life is
developing a theoretical ontology (set of core concepts) able to support fruitful
generalizations (whether deterministic or probabilistic) and models concerning the
possibilities for life. There are no guarantees that the discovery of a biologically
promising, anomalous extraterrestrial phenomenon will yield new insights into the
nature of life; for all we know life may lack a unifying nature. But if it does, and
this is the dream, we may find ourselves in the position of being able to formulate a
more powerful conceptual framework for theorizing about life considered
generally.

As the next chapter (Chapter 9) discusses, we may not need to travel beyond
Earth to discover unfamiliar life. The widespread belief that all life on Earth is
related through a last universal common ancestor is not as obvious as commonly
supposed. Our understanding of the origins of life and the structure and function of
microbial communities is compatible with the idea that there may have been more
than one genesis of life on Earth and moreover that their microbial descendants – in
the form of an as yet undiscovered "shadow biosphere" (Cleland 2007; Cleland
and Copley 2005) – could still be with us today. For the tools used by microbiologists to explore the microbial world (viz., microscopy, cultivation, and metagenomic techniques) would have difficulty detecting even modestly different forms
of microbial life if they exist. The strategy advocated in this chapter for searching
for extraterrestrial life (by identifying and investigating anomalies) also applies to
searching for a shadow biosphere right here on Earth.

9

A Shadow Biosphere: Alien Microbes on Earth?

9.1 Overview

This chapter explores the possibility of a shadow biosphere, that is, a form of microbial Earth life descended from an alternative abiogenesis.[1] It is widely assumed that all life on Earth shares a common origin. Yet there is surprisingly little theoretical or empirical support for this belief, although it is true that all *known* life is so related. As Section 9.2 explains, the possibility that more than one form of life arose on Earth is consistent with (i) prevailing models of the origin of life (the RNA and SM (Small Molecule) Worlds, discussed in Section 5.4) and (ii) our current understanding of molecular biology and geochemical conditions on the early Earth. While the possibility that our planet hosted more than one abiogenesis is often conceded, many scientists nonetheless insist that any descendants would have been eliminated long ago by our microbial ancestors in a Darwinian competition for vital resources. As we shall see, this theoretical argument is undermined by what has been learned in recent years about the structure and dynamics of microbial communities.

The objection dominating the reactions of most scientists to speculation about a shadow biosphere, however, is empirical: We would have found evidence of such microbes if they existed. As astrobiologist Charles Cockell puts it, "I think it is very unlikely that after 300 years of microbiology we would not have detected such organisms despite the fact that they are supposed to have a different biochemistry from the kind we know about today."[2] Section 9.3 explains why this argument is seriously flawed. First, the tools currently used to explore the

---

[1] The term 'shadow biosphere' first appeared in print in Cleland and Copley (2005). It is intended to capture the idea that (like all life) shadow microbes would leave as yet unrecognized traces (in essence, shadows) in their environments as they extract matter and energy and release waste products. The material in this chapter updates and expands upon material presented in Cleland (2007) and Cleland and Copley (2005).

[2] This quotation is from "'Shadow Biosphere' theory gaining scientific support" (Robin McKie, *The Observer*, 2013).

microbial world – microscopy, cultivation, and metagenomic methods – could not detect them, underscoring the old adage that absence of evidence is not evidence of absence. Even more provocatively, as Section 9.4 discusses, a surprising number of anomalies that just might be evidence of an alternative biology have been found. In light of the discussion in Chapter 8, the challenge is convincing researchers to treat them as potential candidates for novel life, as opposed to mere puzzles to be explained in terms of nonliving geochemical processes. There is, of course, no guarantee that these strange phenomena are the product of an unfamiliar biology. On the other hand, when one reflects upon the profound philosophical and scientific importance that the discovery of a truly novel form of life would represent, a dedicated search for such microbes seems well worth the effort.

9.2 How Scientifically Plausible Is a Shadow Biosphere?

In recent years, the conjecture that Earth might be host to an unknown form of microbial life, descended from an alternative origin of life, has generated increasing interest among researchers, for example, Davies et al. (2009), Cleland (2007), Cleland and Copley (2005), Davies and Lineweaver (2005). This is not surprising. As Section 9.2.1 discusses, despite its dazzling morphological diversity, familiar life is remarkably similar at the molecular level. Biochemists have established that the proteins and nucleic acids of familiar life could be at least modestly different in important ways without compromising their biofunctionality.[3] While we do not know the specific chemical and physical processes that gave rise to the earliest life on Earth, we do know that variations in the molecular building blocks of familiar life were available. If the emergence of life is, like other natural phenomena, highly probable under the right chemical and physical conditions, then it seems likely that early Earth hosted alternative origins of life and that some of them gave rise to biochemical variations on life as we know it.

That there may have been variations in the biomolecules of the first living things on Earth is conceded by most molecular biologists. The conjecture that their microbial descendants might still be with us is widely rejected, however, on the following theoretical grounds: (i) Horizontal gene transfer (HGT) would have amalgamated molecular variations among the first proto-organisms into a single form of "advanced" life (capable of evolving by Darwinian evolution); (ii) our

---

[3] The focus of this chapter is on a shadow biosphere composed of microorganisms differing modestly from familiar life in definite ways at the molecular level. The possibility of even stranger forms of life cannot be dismissed but it is very difficult to speculate about what they might be like. Nevertheless, as this chapter intimates, the most promising strategy for hunting for forms of life truly different from our own (whether here on Earth or elsewhere) is to use tentative criteria to identify potentially biological anomalies (along the lines discussed in Section 8.4) for further scientific exploration.

more robust and aggressive microbial ancestors would have long ago eliminated any descendants of an alternative origin of life in a classic Darwinian struggle to survive and reproduce. As Section 9.2.2 explains, neither argument holds up well under close scrutiny. Both reflect a host of assumptions about life that are grounded in observations of complex multicellular eukaryotes – assumptions that are problematic when extended to the world of bacteria and archaea.

9.2.1 Did Life Originate Only Once on Earth?

The morphological diversity of life as we know it is astonishing, ranging from unicellular organisms (bacteria, archaea, and protists, e.g., diatoms) to true multicellular organisms (rotifers, jellyfish, mushrooms, ladybugs, snakes, crows, redwood trees, elephants, etc.). Nevertheless, as discussed in Section 5.2, all of these organisms are extremely similar at the molecular level. A recap of these striking similarities, in the context of what we know about geochemical conditions and processes on the early Earth, is revealing.

Life as we know it uses proteins for the bulk of its structural and catalytic material and nucleic acids (DNA and RNA) to store and manage hereditary information. Proteins are polymers composed of amino acids strung together to form long chains. Although more than 100 amino acids have been found in nature (and over 500 are known from laboratory syntheses), only 20 appear in the genetic code of familiar Earth life. Not just any sequence of amino acids will build a functional protein. The nature and order of amino acids determines (often with the help of molecular chaperones) the three-dimensional structure of a protein, which is crucial to its biological functionality. There must be a sufficient number of large and small, hydrophobic and hydrophilic, and charged amino acids, and they must occur in the correct order. Change the order and you change the functional properties, and hence identity, of the protein. This restricts the collection of amino acids that can be used to build a sufficiently diverse set of proteins for performing the myriad structural and catalytic functions of life. It does not, however, eliminate alternatives to the "standard" 20 amino acids that are coded for in the genomes of familiar life. As Benner (1994) discusses, the number of nonstandard amino acids that could be used to build alternative proteins that would be functional in the right organismal environments is quite large. Moreover, like many organic molecules, amino acids have the property of handedness, which consists of alternative asymmetric arrangements – known as L (levo) and D (dextro) – of chemical bonds around a central carbon atom. The proteins of familiar life (that are encoded directly by the genetic code) are synthesized only from L-amino acids. Because chiral mixtures of L- and D-amino acids do not build good protein structures, it is not surprising that familiar life utilizes amino acids of the same chirality. The mystery is why L- instead of D-amino acids? Proteins that fold correctly, and hence could be functional in the right organismal environments, have been synthe-sized in the laboratory from nonstandard amino acids of both chiralities; see Hong et al. (2014) for a review.

Amino acids, including many which are not used by familiar life, were available on the early Earth for constructing proteins. Empirical and theoretical investiga-tions have revealed a diversity of mechanisms for producing nonstandard amino acids of mixed L- and D-chirality under conditions thought to have been present on the primordial Earth. These conditions include physicochemical events involving the Earth's atmosphere – such as electrical discharges (mimicking lightning) through mixtures of simple gases (e.g., Jiang et al. 2014; Miller 1953, 1955), frequent solar shock waves from a stormy young sun (e.g., Airapetian et al. 2017), and photochemical reactions inside atmospheric aerosols (Dobson et al. 2000) – hydrothermal processes (such as those found in oceanic volcanic vents) (Hennet et al. 1992; Holm and Andersson 1998; Martin and Russell 2003), and geochem-ical processes involving porous mineral surfaces (Cairns-Smith et al. 1992; Cody 2004; Wächtershäuser 1988). In addition, more than 80 amino acids have been found in carbonaceous meteorites (Glavin et al. 2006). While some meteorites contain an excess of L- over D-amino acids (Andersen and Haack 2005), D-amino acids are nonetheless common. It would thus be very surprising if the chemical resources for building proteins were the same in every incipient "cradle of life" on the young Earth. This suggests at least three plausible ways in which the proteins of the first living things on Earth could have differed modestly from familiar life: (1) They might have been built from the same 20 amino acids as familiar life but of the opposite (D) chirality, (2) they might have been built from a set of L-amino acids containing some nonstandard amino acids, or (3) they might have been built from a set of D-amino acids containing some nonstandard amino acids. Accordingly, on metabolism-first accounts of the origin of life, such as the SM (Small Molecule) model discussed in Section 5.4, it seems likely that the early Earth hosted more than one abiogenesis; for a variety of nonstandard building blocks for a proto-metabolic system (collectively autocatalytic, chemical reaction network) were available.

The complexity of the nucleic acid-based hereditary system of familiar life is greater than that of its protein-based metabolic system, suggesting additional ways in which life might differ at the molecular level. Nucleic acids are polymers made of monomeric nucleotides strung together in long chains. A nucleotide consists of three molecular subunits, a sugar, a phosphate, and a nucleobase. DNA and RNA use different sugars – RNA uses ribose and DNA uses deoxyribose – but are of the same (D) chirality. Hereditary information is encoded by five (nucleo)bases: adenine (A), guanine (G), cytosine (C), thymine (T), and uracil (U). A, G, C, and T are used to encode hereditary information on DNA. RNA uses the same bases except that it substitutes U for T. Bases pair off in a complementary pattern, with A pairing with T (U in RNA) and G pairing with C. Hereditary information is encoded on a single (the coding) strand of duplex DNA by sequences of three consecutive bases. Each triplet of bases ("codon") corresponds to a specific amino acid or the initiation or termination of a chain of amino acids constituting a protein. The process of translating information encoded on DNA into protein begins with the construction of a complementary strand of messenger RNA (substituting U for T in the process). The latter is transported to a ribosome, which decodes the information and translates it into protein.

Modest variations in the molecular composition and architecture of our DNA–RNA based hereditary system have been explored both theoretically and in the laboratory. Life could have used different nucleobases to encode hereditary information. A variety of standard and nonstandard bases have been discovered in carbonaceous meteorites (Callahan et al. 2011) and synthesized in the laboratory under conditions thought to have been present on the early Earth (e.g., Powner et al. 2009). In addition, life could have used different sugars in the sugar–phosphate backbones of DNA and RNA. Standard (deoxyribose and ribose) and nonstandard sugars of different chiralities can be synthesized from formaldehyde using the formose and related reactions (Gesteland and Atkins 1993). Alternatives to the genetic code of familiar life are also possible. Although no triplet of bases is paired with more than one amino acid, some amino acids (e.g., serine) are paired with more than one triplet of bases (TCT, TCC, TCA, TCG), suggesting flexibility in assignments of codons to amino acids. Moreover, there is little reason to suppose that life could not use a doublet, quadruplet or even quintuplet (versus triplet) coding scheme. While a triplet coding scheme is the most efficient for four bases and 20 amino acids, since $4^3 = 64$ possible distinct coding sequences, it would not be optimal for different numbers of bases and amino acids.

It is one thing to demonstrate that the early Earth is likely to have contained the building blocks for nonstandard nucleic acids but quite another to establish that a short nucleic acid molecule (oligomer) could be synthesized from these building blocks under prebiotic conditions. Biochemists still do not know how standard nucleic acids, let alone the complex cooperative two-polymer (nucleic acid–protein) system of familiar life, could arise under natural conditions. As Section 5.4 discussed, the RNA World currently dominates scientific thought about the origin of life.[4] But even supposing that the basic molecular building blocks of RNA – most notably, sugars and bases, since phosphate is abundant – were present in

---

[4] Unlike DNA, RNA can catalyze itself, which reputedly solves a "which came first, the chicken or the egg problem" for the origin of life; RNA can perform the catalytic function of proteins and store hereditary information like DNA. RNA also plays the central role in translating genetic information into proteins in ribosomes.

sufficient concentrations on the early Earth,[5] the chemical reactions required to synthesize RNA oligomers are thermodynamically unfavorable, especially in aquatic environments, which are widely viewed as the most likely sites for the emergence of life. As Cafferty and Hud (2014) discuss, in a review of challenges for the RNA World, sugars do not react with bases to form nucleosides under plausible prebiotic conditions, and even supposing that this problem could be overcome, there is the additional difficulty of getting a nucleoside to bond with a phosphate unit to form a nucleotide, and the further problem of polymerizing a pool of nucleotides into an RNA oligomer capable of catalyzing itself, or alternatively, a collection of diverse RNA molecules capable of collectively catalyzing each other. Nevertheless, from a biochemical standpoint, there is no reason to suppose that synthesizing RNA oligomers from nonstandard bases and sugars is any more challenging geochemically than synthesizing an RNA oligomer from standard bases and sugars. Both seem equally implausible from a geochemical perspective.

In the face of these difficulties with making good geochemical sense of the emergence of even a standard RNA oligomer from chemical compounds available on the early Earth, some biochemists follow Francis Crick (1981) in contending that the origin of the first RNA oligomer was an extremely improbable (one-off) event, in essence a "scientific miracle." On this view, the emergence of a nonstandard RNA World is even more improbable than that of the standard RNA World since it requires two independent miracles, one producing a self-replicating, standard RNA oligomer and another producing a self-replicating, nonstandard RNA oligomer; there is the additional difficulty that most RNA molecules do not make good catalysts. Even worse for the scientific miracle argument, most contemporary models of the RNA World presuppose the abiotic synthesis of a collection of mutually catalytic RNA oligomers (as opposed to a self-catalytic grandmother RNA oligomer), which requires multiple miracles occurring contemporaneously in the same location just to generate the standard RNA World. So many improbable events occurring together in the same locale renders the probability of even familiar life arising on the early Earth via a standard RNA World extremely low – so low as to be almost as improbable as my passing through a solid metal door, which according to quantum mechanics is (from a mathematical perspective) possible but so improbable as to be physically impossible.

A proponent of the RNA World need not, however, endorse the radical position of Crick and fellow travelers regarding the natural origin of the first RNA

---

[5] Concentration is a big problem: Bases are not found in large numbers in meteorites, nor are they synthesized in large quantities under laboratory conditions thought to represent those found on early Earth. Sugars are unstable, which suggests that even supposing that they were readily synthesized on early Earth, they are unlikely to have accumulated in large enough quantities.

oligomer(s). The *only* evidence for the cosmic coincidence conjecture is ignorance about how an RNA oligomer could arise under plausible prebiotic conditions on the early Earth. To the extent that science operates under the guiding principle that natural phenomena are explicable in terms of specifiable natural processes, appeals to cosmic coincidence to "explain" (the term is really an oxymoron) the origin of RNA places the RNA World beyond the reach of scientific investigation just as surely as an appeal to supernatural creation; see Iris Fry (1995) for a general discussion of this issue. Besides, our ignorance about how RNA could arise on the young Earth is equally compatible with several more realistic conjectures, including: (i) Scientists do not fully understand the relevant physical and chemical conditions and processes that gave rise to the first RNA oligomers on Earth; (ii) critical stages in the synthesis of RNA require exotic conditions and processes – for example, found in interstellar clouds (Belloche et al. 2014; Chatterjee 2016; Ehrenfreund and Menten 2002) – that were never present on Earth; (iii) the RNA World replaced an earlier informational polymer that was easier to synthesize under prebiotic conditions (e.g., Cech 2012; Nelson et al. 2000; Orgel 1998, 2000). Most importantly, for our purposes, these speculative theoretical mechanisms are neutral with regard to the production of standard or nonstandard RNA (or their respective building blocks). The upshot is that if the emergence of an RNA World is, like other natural phenomena, highly probable under certain (albeit as yet unknown) physicochemical conditions, then it is not at all implausible that the early Earth "experimented" with modest variations on standard RNA.

To wrap up, we know that both nonstandard and standard building blocks for proteins and nucleic acids were present on the early Earth. We do not know how the first proteins and nucleic acids were synthesized from them. We do know that, given our current understanding, generating a primitive metabolic system (SM World) or "ribo-organism" (RNA World) from nonstandard molecular building blocks for (respectively) proteins and nucleic acids does not seem any more challenging than generating one from standard molecular building blocks. Perhaps the SM World and the RNA World models are both badly mistaken, in which case, all bets are off. But assuming that one of these models is on the right track, there is little reason to suppose that the early Earth could not have hosted modestly different forms of primitive life.

9.2.2 Could the Present-Day Earth Be Host to a Shadow Biosphere?

It is clear that there are no large multicellular descendants of an alternative biogenesis on Earth today. But the lack of such organisms should not count against the plausibility of a *microbial* shadow biosphere. There are compelling reasons for suspecting that the most common form of life in the universe is unicellular and that complex multicellular organisms are very rare (Ward and Brownlee 2000). Geochemical studies of carbon isotope ratios found in ancient rocks suggest that there was microbial life on Earth more than 3.8 bya (Mojzsis et al. 1996) and perhaps as far back as 4.1 bya (Bell et al. 2015), which (geologically speaking) is shortly after its formation around 4.5 bya. The metabolic processes of familiar life exhibit a pronounced preference for the lighter stable isotope of carbon (^{12}C) over the heavier stable isotope (^{13}C). This preference is difficult to explain in terms of known abiotic processes. But even supposing that these, somewhat controversial, geochemical findings are mistaken, studies of stromatolites (fossilized microbial mats) indicate that life was well established on Earth by 3.5 bya. Evidence of "true" multicellular organisms (highly specialized, functionally integrated systems of cells[6]) does not appear in the geological record until fairly recently, around 640 mya; more primitive multicellular eukaryotes may go back as far as 2.1 bya (El Albani et al. 2010), but multicellular animals are (geologically speaking) a recent phenomenon. In short, life on Earth was unicellular for at least 2–3 billion years before giving rise to complex multicellular organisms less than a billion years ago. The antiquity of unicellular microbes vis-à-vis complex multicellular eukaryotes suggests that the evolution of complex multicellular organisms from unicellular microbes is the exception rather than the rule.[7] It follows that the lack of complex multicellular organisms descended from a separate origin of life has little bearing on the question of whether a microbial shadow biosphere exists on Earth today.

Skeptics rarely, however, cite the absence of molecularly deviant macroscopic organisms as grounds for rejecting the plausibility of a shadow biosphere. Instead, they contend that any differences among the primitive precursors to modern cells would have been amalgamated (via HGT) into a single form of cellular life or, alternatively, any unicellular microorganisms descended from an alternative biogenesis would have been eliminated long ago in a Darwinian competition for vital resources with our more robust and aggressive unicellular ancestors. As discussed below, neither argument holds up under close scrutiny.

The phenomenon of horizontal gene transfer (HGT) – also known as lateral gene transfer (LGT) – is sometimes held up as precluding the possibility of

---

[6] Bacteria and archaea sometimes form multicellular colonies in which each cell can exist on its own (carry out all the functions of life) but benefits from a close association with other cells in a group of cells for purposes of defense, reproduction, or to attack larger prey. But the cells in such colonies lack the specialization and functional integration (into tissues and organs) found in true multicellular organisms, such as plants and animals, whose cells cannot exist on their own.

[7] The difficulties with evolving complex multicellular life under natural conditions could provide a partial answer to the famed Fermi Paradox: Where are all the technologically intelligent extraterrestrials? The answer being intimated is that complex multicellular life, which is what eventually gives rise to intelligent life, is extraordinarily rare in the universe, and intelligent life rarer still; and if the latter becomes technologically sophisticated, then (if the current predicament of humanity provides a guideline) it usually destroys itself.

shadow microbes. As Section 6.2 discussed, HGT occurs when genes are transferred directly between unrelated organisms, as opposed to being transferred vertically from parent to offspring in the process of reproduction. The three main mechanisms for HGT are (1) uptake and incorporation of fragments of foreign DNA – typically released into the environment by dead cells – into a recipient cell's genome (transformation), (2) direct exchange of DNA through cell-to-cell contact (conjugation), and (3) insertion of genetic material directly into a host cell's genome as a result of viral infection (transduction). All three mechanisms are common among prokaryotic microbes, providing them with a source of heritable variation other than mutation. Eukaryotes, on the other hand, have barriers to HGT. Their chromosomes are encased within a protective nuclear membrane, which shields them from the first two mechanisms of HGT. In addition, true multicellular eukaryotes have specialized germ cells for reproduction which decreases the likelihood that foreign DNA, acquired via viral infection (transduction), will be inherited by their offspring. Nevertheless, HGT occurs across all three domains of life (Archaea, Bacteria, and Eukarya). Approximately 5–10% of mammalian genes have been acquired via transduction from retroviruses, and some of them are thought to have played key roles in the evolution of mammals. Retroviral genes, for instance, are thought to have enabled the evolution of the mammalian placenta (Dupressoir et al. 2012).

Many biologists believe that HGT played a central role in the evolution of modern microbes from their primitive precursors. Carl Woese, who has written more extensively on this topic than anyone else, conjectures that the first cell-like entities ("progenotes") were very different from modern unicellular microbes. They were not genuine organisms, that is, "self-replicating entities that have true individuality and a history of their own" (Woese 1998, 6857). According to Woese, a proponent of the RNA World, the ancestor of familiar life is not, as sometimes portrayed, a single lucky cell but instead a loosely confederated community of (what amounts to) tiny "bags" of precursor biomolecules (especially RNA oligomers and their building blocks, but also amino acids, small peptides, and co-factors), subject to extensive mutation and rampant "HGT," reshuffling their contents as they split and merge in haphazard ways. Eventually, through a process of trial and error, the complex cooperative arrangement of proteins and nucleic acids (that underlies the phenotype–genotype distinction of familiar life) appeared, and the first free-living cells (modern unicellular organisms), capable of evolving by means of Darwinian evolution, emerged. On Woese's scenario, progenotes containing nonstandard and standard biomolecules would have been amalgamated by HGT into a single form of life *before* the emergence of unicellular microorganisms.

This argument fares little better than the earlier one. The "HGT" used by Woese's progenotes bears little resemblance to the mechanisms of HGT used by modern cells. Progenotes freely swap all kinds of cellular componentry (Woese

1998, 6856). HGT among modern unicellular organisms, in contrast, is highly selective, involving not only the transfer but also the integration of segments of DNA or RNA into another cell's genetic system. The latter presupposes significant similarities in the genetic machinery for replication, transcription, and translation, which helps to explain why HGT is most common among microbes from the same domain of life and rarer among microbes from different domains. In truth, cross-domain HGT is possible only because the genetic machinery of organisms from all three domains is very similar at the molecular level. No microbe from any of the three domains of life could integrate genomic sequences using nonstandard nucleobases into its genome. And even supposing that such an event occurred, the alien gene could not be replicated or expressed (translated into protein by ribosomes).

According to Woese (2004), it was not until a lucky progenote hit upon the right combination of biomolecules for a modern genetic system that it crossed the "Darwinian threshold" and produced a true biological individual (a modern unicellular organism) containing the highly sophisticated genetic machinery required for modern HGT. Moreover, on his account, the three domains of familiar life achieved this transition separately, with bacteria crossing the Darwinian threshold first, followed by archaea, and finally unicellular eukarya. Granting for the sake of argument that Woese is right about this, could some of these communitarian progenotes have produced even more deviant forms of microbial life (differing in their molecular composition or architecture in some of the ways discussed in Section 9.2.1) that crossed the Darwinian threshold independently of familiar unicellular microbes? Nothing in his account rules out progenotes containing biomolecules (nucleic acids and proteins) built from nonstandard molecular precursors, especially if geographical isolation occurred. Furthermore, there is no reason to suppose that such progenotes would have more difficulty in producing cells capable of crossing the proverbial Darwinian threshold than those that allegedly gave rise to the three domains of familiar cellular life. Admittedly, this argument is highly speculative. But so is Woese's account of the origin of the three domains of familiar life. The point is there is no reason to suppose that Woese's progenotes could not have given rise to an alien form of cellular life; many biologists do not, however, accept Woese's account of the origins of the three domains of life.

I am not claiming that a (microbial) shadow biosphere exists on Earth today. I am merely pointing out that the theoretical argument from HGT against the plausibility of such microorganisms is not as compelling as often claimed by its proponents. The same goes for the argument that familiar microbes would have long ago outcompeted any unicellular descendants of an alternative origin of life. As discussed below, the latter line of reasoning is undermined by what microbiologists have learned in recent years about the structure and functional dynamics of microbial communities.

The proverbial Darwinian account of life as "red in tooth and claw" is based upon observations of large, complex, multicellular eukaryotes, especially animals. Interactions among the inhabitants of microbial communities, in contrast, are more "collaborative" than aggressive (Dupré and O'Malley 2009; O'Malley 2014, 9–10), a point highlighted by John Dupré's provocative reference to "the friendly germ" in his 2007 Horning Lecture (Oregon State, Dupré 2007). Microbial communities are complex ecosystems, consisting of functionally integrated groups of microorganisms. Members of these communities modify their environments chemically and physically in ways that help to sustain the community as a whole, creating a diverse collection of fairly stable microenvironments that would not otherwise exist. While typically dominated by a few, highly abundant, varieties of microbes, the bulk of the phylogenetic diversity of most microbial communities consists of tiny populations of "rare" microbes (Magurran and Henderson 2003; Sogin et al. 2006). Being present in low numbers is not an evolutionary disadvantage for a rare microbe as it is in the case of animals. When populations of mammals drop below a certain threshold they become endangered by more abundant populations of competitors or predators. Small populations of rare microbes, in contrast, are protected by occupying specialized ecological niches not occupied by common microbes, more specifically, by producing or utilizing material that is utilized, produced or ignored by other varieties of microbe. Depending upon its nature, environmental degradation of an ecosystem may have a more deleterious effect on a common variety of microbe than on a rare one (Kirchman 2012, Ch. 9).

Most microbial communities include microorganisms from all three domains of life, as well as viruses, whose status as living is controversial (see Section 6.4). When first "discovered," archaea were regarded as extremophiles found only in very hot, salty, and acidic environments. Archaea have since been found coexisting with bacteria under a wide range of moderate conditions (e.g., soils and ocean surface waters) and bacteria have been found coexisting with archaea under extreme conditions. As Kirchman (2012, Ch. 9) discusses, the distribution of archaea vis-à-vis bacteria depends upon the nature of the environment, and this variation is best understood in terms of their respective physiology. Most bacteria exploit either organic matter or light as energy sources, whereas most archaea exploit inorganic chemicals. It is thus not surprising that archaea are more numerous in environments less hospitable to bacteria, such as the deep (dark) ocean or high temperature hot springs, and that bacteria are more numerous in ocean surface waters and soils in temperate regions.

The relative distributions of archaea and bacteria across different environments suggests two evolutionary strategies that shadow microbes might exploit in coping with familiar microbes. They might be present in many microbial communities at low levels, exploiting energy resources that are underutilized or ignored by archaea and bacteria. An especially intriguing possibility is harvesting energy from gravity (versus light or chemicals) through pressure gradients in, for instance, tidal zones (Schulze-Makuch and Irwin 2008). Alternatively, shadow microbes might create their own integrated ecosystem in an environment containing little or no familiar life, such as extremely dry deserts, high latitude ice sheets, high altitude mountain tops, the middle to upper atmosphere, ocean vents at temperatures above 130 °C, or deep beneath the Earth's crust (Davies et al. 2009). Ultraviolet light, which is plentiful at high altitudes but lethal to familiar life, provides an intriguing potential energy source for an alternative form of microbial life.

9.3 If They Exist, Why Have We Not Found Them?

Perhaps the most compelling objection to a microbial shadow biosphere is empirical: If they exist we would have discovered them by now, or at least encountered telling signs of them. This objection ignores limitations of current technologies (microscopy, cultivation, and molecular biology techniques) used to explore the microbial world. As discussed below, none of these tools could detect microbes differing even modestly from familiar life in their basic molecular composition or core molecular architecture.

9.3.1 Limitations to Microscopy

The utility of microscopy for identifying truly novel microbes is exceedingly limited. Molecular biology has taught us that superficial similarities in gross cellular morphology can hide important differences at the molecular level. Archaea provide a case in point. They share the prokaryotic cell plan with bacteria, but (as discussed in Section 6.2) differ genetically and biochemically from bacteria more than they do from eukarya. Shadow microbes with a prokaryotic cell plan could not be discriminated from bacteria under a microscope any more easily than archaea. The inability to discriminate bacteria from shadow microbes is not a minor worry. As is well known, evolutionary pressures sometimes produce similar adaptations from different biological building blocks. Insects, birds, and bats, which do not share a common ancestor with wings, provide a good illustration. Just as conditions on the later Earth favored the independent development of wings in insects, birds and bats, so conditions on the early Earth may have favored the development of a prokaryotic cell plan, which could explain its somewhat surprising lack of phylogenetic significance for bacteria and archaea. In short, one cannot conclude, as is commonly supposed, that every microbe having a prokaryote cell plan is either an archaeon or a bacterium. It is possible that we have already viewed shadow microbes under a microscope and, deceived by their gross cellular morphology, failed to recognize what they represent.

Our ability to visualize microbes under a microscope has been extended in recent years by the development of specialized molecular staining techniques for viewing particular cellular components, including, most importantly, biomolecules such as nucleic acids and lipids. These stains are unlikely to be very useful in discriminating shadow microbes from morphologically similar familiar microbes, however. Some are too coarse grained to differentiate familiar microbes from shadow microbes. DAPI (4′,6-diamidino-2-phenylindole), a stain widely used for detecting microbes in environmental samples, provides a good illustration. It intercalates into double stranded DNA in living cells, causing it to fluoresce, and would almost certainly stain duplex DNA using nonstandard nucleobases. Hybridization techniques, such as FISH (fluorescence in situ hybridization), on the other hand, are too specific to the DNA of familiar life to detect shadow microbes. FISH causes specific regions of chromosomes within intact cells to fluoresce by hybridizing them with a fluorescently labeled complementary oligonucleotide probe composed of standard bases. The utility of FISH is thus limited to cells whose DNA base sequences are complementary to the probe being used. DNA consisting of nonstandard bases would not hybridize with the probe, and hence would not fluoresce under a microscope.

9.3.2 Limitations to Cultivation

Our most detailed knowledge of microbes has been achieved through cultivation, that is, by cloning (growing many copies of) a single microbe under controlled laboratory conditions. Cultivation yields large quantities of essentially identical microbes, providing biologists with enough material to perform extensive analyses of their genetic, structural, and enzymatic composition.

Our ability to cultivate microbes is exceedingly limited, however. It is estimated that cultivation-based methods find less than 1% of bacteria and archaea (Pace 1997). That the vast majority of prokaryotes cannot be cultivated is somewhat surprising when one considers that many of these missing microbes coexist with microorganisms that can be cultivated; one would think that they would thrive under similar conditions. A plausible explanation is that the uncultivated microbes are dependent on complex, poorly understood, biogeochemical conditions created by microbial communities. On the other hand, many uncultivated microorganisms thrive under extreme physical and chemical conditions (especially of pressure, temperatures, and pH) and exploit an astonishing range of inorganic material (e.g., hydrogen sulfide, ferrous iron, manganese, and ammonia) as nutrient sources.

Because it is difficult to fully identify let alone replicate these conditions in a laboratory setting, it is hardly surprising that it is difficult to cultivate them. When one considers trying to cultivate shadow microbes, which are even more likely to require unanticipated growth conditions, these difficulties are exacerbated. Indeed, even supposing that shadow microbes coexist with familiar microbes in microbial ecosystems, they are likely to have nonstandard nutrient requirements (even under "normal" conditions). Consequently, the fact that no colonies of biochemically deviant microbes have been found growing in Petri dishes does not count much against the existence of microbes descended from an alternative origin of life.

9.3.3 Limitations to Metagenomic Methods

The most powerful techniques for exploring the microbial world are culture-independent metagenomic methods. These methods extract DNA or RNA directly from environmental samples, circumventing the need for isolating and culturing wild microbes, and generate enough copies for genomic analysis. The earliest culture-independent studies of microbial diversity used PCR (polymerase chain reaction) to target and amplify (clone) specific genes, namely, those coding for 16s rRNA in bacteria and archaea and for 18s rRNA in eukarya.[8] PCR amplification of DNA requires "primers" (small pieces of synthetic DNA) to delimit the genomic sequence to be copied, which limits the capacity of PCR to "recognize" wild genomic sequences. Because universal primers for bacteria could not pick out the distinctive 16s rRNA genes of archaea, the first PCR studies of microbial diversity missed them. Subsequent studies, using primers for all three domains of life, revealed that the vast majority (around 99%) of bacteria and archaea had been missed by cultivation (Pace 1997). The important point for our purposes, however, is that PCR presupposes prior knowledge of a genomic sequence if it is to be able to select it for amplification. No primer based on any of the three domains of familiar life would be able to recognize genomic sequences from an alternative form of microbial life. Moreover, the focus of PCR studies of microbial diversity on propagating ribosomal genes means that they could not detect microbes lacking ribosomes even supposing that they used standard nucleic acids. It is highly unlikely that the first proto-organisms on Earth used anything as sophisticated as ribosomes (minuscule molecular machines composed of both RNA and protein), which means that PCR studies could not even detect our primitive progenitors; some biochemists even speculate that primitive riboorganisms might still be emerging on Earth.

---

[8] These genomic sequences were selected for study because they have regions with different levels of variability. Some regions are highly conserved (virtually unchanged by evolution down through geological time) whereas others are highly variable. They thus provide good tools for mapping phylogenetic relations among microbes.

First-generation "shotgun sequencing" improved upon early PCR approaches by allowing microbiologists to (at least in principle) propagate and sequence all the genes (protein as well as ribosomal) in an environmental sample. DNA in the sample is cut into short segments and hybridized with "vectors" (small units of DNA derived from a virus or bacterial plasmid). The recombinant DNA is inserted into a host microorganism, typically an *E. coli* bacterium, which is then cultured, producing many copies of the foreign DNA along with the host microbe. The foreign gene fragment is sequenced, compared with a "library" of other gene fragments (extracted from the sample) for overlap, and pieced together into longer genomic segments, including whole chromosomes.

Shotgun sequencing revealed that PCR-based surveys were missing large numbers of genetically divergent, low abundance, archaean microbes (Baker et al. 2006), suggesting that our understanding of the diversity of microbial life on Earth is still very limited. Next-generation (also known as high-throughput) sequencing methods confirmed this suspicion, recovering evidence of even stranger microbes from environmental samples. Indeed, some of the gene fragments discovered by next-generation methods are so different from those of any known archaeon or bacterium, not to mention eukaryan, that it is difficult to classify them taxonomically in terms of the standard domains of life (Yooseph et al. 2007). A few researchers (e.g., Wu et al. 2011) go so far as to speculate that they may belong to new domains of life. Most microbiologists, however, are highly skeptical, contending that they are either viral in origin or artifacts of some sort. Somewhat intriguingly, the argument that they are viral is based on the absence of a signature subunit of rRNA (Moreira and López-García 2015), reflecting the assumption that all unicellular microorganisms have ribosomes; viruses lack them.[9] As discussed in the previous chapter (Chapter 8), this reaction is not surprising. Anomalies – and these strange genomic sequences are anomalous, despite the fact that they use standard nucleic acids – are usually treated as explicable in terms of widely accepted scientific commitments, in this case, the conviction that the three domains of life are exhaustive.[10] The discovery of another domain of familiar life would not, of course, be the same as discovering a shadow biosphere, that is, a truly novel form of life descended from an alternative origin. The gene fragments concerned are potentially biologically anomalous only vis-à-vis the assumption that there are just three domains of familiar life.

---

[9] A few researchers contend that giant viruses should be classified as living things and placed in their own domain of life, but this is not a widely shared view.

[10] As discussed in Section 6.4, they may well be right. Most puzzling empirical findings are eventually resolved within the context of widely accepted scientific beliefs. Nevertheless, those that prove especially difficult to explain in terms of current scientific beliefs often do give rise to major scientific discoveries.

It is clear that none of the newer genomic methods for exploring the microbial world could detect a truly alternative form of microbial life. First-generation shotgun sequencing requires combining genomic material recovered from the environment with a vector, inserting the recombinant DNA into a host microorganism, and cloning it. Even supposing that a genomic fragment from a shadow microbe could hybridize with a vector, it is highly unlikely that the genetic machinery of the host microorganism could replicate it. Next-generation sequencing technologies enable researchers to sequence enormous quantities of environmental DNA in parallel, allowing them to detect low abundance varieties of microbes missed by first-generation methods. These methods by-pass the need for a host by attaching wild DNA fragments to solid surfaces (in the Illumine sequencer, the surface is beaded and the fragments are anchored to the beads), allowing millions of sequencing reactions to happen concurrently. The processes involved use small bits of artificial DNA (adaptors and primers), however, which means that they cannot sequence gene fragments containing nonstandard bases.

In sum, all the genomic methods currently available for investigating microbial diversity presuppose that the wild microbe is (at most) an exotic form of familiar life. It is thus hardly surprising that metagenomic methods have not found evidence of a genuinely alternative form of microbial life. In truth, it is difficult to design an effective metagenomic method, using bits of artificial nonstandard DNA (primers and adaptors), for detecting shadow microbes in advance of knowing what to look for. But the fact that no one has even tried to do so reflects the widely shared theoretical assumption, guiding empirical studies, that all life on Earth descends from a common ancestor. That microbiologists keep finding increasingly divergent varieties of microbes (as their methods for exploring the microbial world become more refined) is very suggestive, however, raising the question: What else is out there waiting to be found?

Metagenomic studies have taught us a lot about microbial ecosystems. Some of these discoveries suggest strategies for searching for a shadow biosphere. The greatest microbial diversity is found in the oceans, whereas the least diversity is found in extreme physical and chemical conditions, especially of temperature, salinity, and pH (Kirchman 2012, Ch. 9). Desert saltpans with very high salt concentrations, for instance, host only a few types of bacteria and archaea. The same is true of acid ponds polluted by runoff from mines and hot springs at high temperatures. Another, perhaps even more promising, place to look is in extreme environments containing few microbes of any type. Some samples of Antarctic soil contain so few microbes that the GCMS (gas chromatograph mass spectrometer) included in the Viking biology instrument package would not be able to detect them (Section 8.2). The upper atmosphere is exposed to high levels of ultraviolet light which, as mentioned earlier, is lethal to familiar life. Environments with biologically unexploited energy sources that are inhospitable or even lethal to familiar life are especially good places to search for shadow microbes. They would stand out against the backdrop of nonliving geochemical processes much more readily than in environments teaming with familiar life. The anomalous character of archaea was first noticed under analogous circumstances, although as discussed in Section 5.3, the puzzling features in question were initially chalked up to convergent evolution in extreme environments.

9.4 Potentially Biological Anomalies: Have We Already Encountered Them?

Given that we really do not know what to look for, the most promising strategy for searching for truly alien life is to seek out potentially biological anomalies (versus life *per se*) using tentative (versus defining) criteria. The approach recommended in Chapter 8 is to screen for structural, compositional, and functional characteristics, some of which are suggestive of biology and others seemingly incompatible with it. As discussed in the last section, environments containing little or no familiar life are good places to begin. The problem is that (*if* they exist) shadow microbes and familiar microbes arose and evolved under very similar planetary conditions (viz., on the young Earth) and hence are likely to resemble each other in ways that make it difficult to discriminate shadow microbes from familiar microbes. As we have seen, microbes differing modestly from familiar life in some of the building blocks of their basic biomolecules could not be detected using contemporary metagenomic methods. Nevertheless, an environment harboring shadow microbes that was lethal to familiar life would likely register as anomalous in a provocative way. It would exhibit a suite of characteristics that "should not be there" (under the assumption that all the processes involved are abiotic). The advantage of using a suite of tentative (versus defining) criteria is that it is difficult to say in advance what these telltale characteristics might be; they might include unusual carbon-based polymers enclosed by strange, membrane-like, mineral structures, for instance. The chief worry, however, is that environments lethal to familiar microbes would also be lethal to shadow microbes. Indeed, given what we know about the distribution of archaea vis-à-vis bacteria, I suspect that (if they exist) shadow microbes coexist with familiar microbes but are most abundant in environments with fewer archaea and bacteria and least abundant in environments swarming with bacteria or archaea. That shadow microbes are likely to coexist with familiar microbes raises the difficulty mentioned earlier of recognizing them against the noisy backdrop of familiar life.

The good news is that scientists have already encountered some intriguing *potential* candidates for an unfamiliar biology but have failed to recognize them

(a)

(b)

Figure 9.1 (a) Desert varnish (dark stripes) streaking a rock wall at Grand Gulch of the Cedar Mesa Plateau (Utah, USA). (Photo courtesy of Cara Marie Lauria.) (b) Close-up of varnish coating on a small piece of red sandstone. (Photo courtesy of Cara Marie Lauria.)

as such. In keeping with the usual scientific response to anomalies, discussed in Section 8.3, these phenomena are acknowledged as genuinely puzzling (resisting efforts at explanation) but nonetheless treated as the products of poorly understood familiar (biological or geochemical) processes. This section focuses on three cases that are especially intriguing in virtue of exhibiting a baffling mixture of allegedly biological and abiotic characteristics: (1) desert varnish, (2) the mysterious "nanobes," described by Philippa Uwins and colleagues (1998), and (3) the strange "nanobacteria," whose status as living entities remains very controversial.

Desert varnish is a mysterious dark coating found on rocks in arid regions (Figure 9.1). Although it has been studied extensively by geologists for more than a hundred years,[11] its origin remains controversial. Consisting of extremely thin chemical and mineralogical layers, desert varnish bears an uncanny morphological resemblance to stromatolites, microbial mats whose fossilized remains date back 3.5 billion years; they are still found today, most famously, in Sharks Bay, Australia. Even more provocatively, varnish coatings are enriched in manganese and iron, which gives them their dark color, despite the fact that the rocks on which they are found are not. A mysterious process is extracting manganese and iron from the environment and concentrating it in the rock coatings.[12] To many geologists this intriguing combination of microstructural and chemical features

---

[11] Darwin and the nineteenth century naturalist and explorer Alexander Humbolt were intrigued by desert varnish; Darwin (1871, 12–13) even speculated that it was the product of biological processes.

[12] Native Americans from the southwest region of the USA exploited the contrast between the dark coatings and the lighter rock beneath in their beautiful petroglyphs.

suggests a biological origin, especially since bacteria and algae commonly produce manganese or iron as byproducts of metabolism.

Other features of desert varnish, however, seem incompatible with a biological origin. Living stromatolites are formed in shallow water by photosynthetic bacteria, which trap and cement sedimentary particles into layers as they grow outwards towards the sunlight. The surfaces of living stromatolites are thus covered in living bacteria. Although fragments of microbial cells, including amino acids and nucleic acid oligomers, are entombed in varnish coatings, few living microbes are found on their surfaces (Perry et al. 2006). Moreover, PCR analysis of 16s rRNA genes extracted from varnish coatings and surrounding rocks and soils reveals variation from region to region, suggesting that no one group of bacteria, let alone a single variety, is responsible for the coatings (Perry et al. 2002). Another feature of desert varnish that seems incompatible with biology is the extreme slowness (over a period of more than 2000 years) with which the coatings form (Liu and Broecker 2000). Finally, attempts to produce varnish-like coatings in laboratory settings using a variety of bacteria and algae have been unsuccessful.

To geologist Randall Perry, who has written extensively on the topic, these features suggest that desert varnish is produced by nonliving processes. Perry et al. (2006) theorize that it is slowly formed through a complex series of geochemical processes involving interactions between wind, water, and rock. The problem is that they have not been able to explain the large amounts of manganese and iron found in the coatings. Yet of all the puzzling characteristics of desert varnish, the presence of inexplicably large amounts of manganese and iron is the most suggestive of biology. In short, desert varnish qualifies as a potentially biological anomaly: It both resembles and fails to resemble familiar life in provocative and unanticipated ways, and moreover has resisted persistent efforts, over a period of more than 100 years, to explain it in terms of known biological or abiotic processes. Of course, none of this guarantees that desert varnish is the product of a novel form of Earth life.

Nanobes and nanobacteria, which closely resemble each other, are also anomalous in the pertinent way. The term 'nanobe' was coined by Philippa Uwins, who found tiny filamentous structures shaped like bacteria growing on rock cores retrieved from Jurassic and Triassic sandstones off the west coast of Australia (Uwins et al. 1998). These structures were composed of the so-called "biological" elements, most notably carbon, oxygen, and nitrogen. Ultrathin sections viewed under a transmission electron microscope revealed an outer membranous layer surrounding an inner area that contains (like familiar unicellular microbes) regions of different densities. The smallest nanobes are just 20 nm in diameter, however, making them far smaller than any known cellular organisms; the smallest known bacterium, *Mycoplasma gallicepticum*, is 200 nm wide. In the view of many biologists, the small size of nanobes cinches the case against life: They are far too small to contain the genetic and metabolic machinery required by [familiar] life. Yet Uwins's nanobes reacted to three DNA stains, including DAPI, which strongly suggests that they contain DNA. Attempts by Uwins and her colleagues to extract and sequence this DNA were unsuccessful, however (Davies 2010, Ch. 3). Despite the fact that one would not expect to be able to isolate and sequence DNA from microbes using a different suite of nucleobases in their DNA, the scientific community quickly lost interest in investigating them further. Whether these strange structures are living things, as opposed to unusual geochemical structures, or even artifacts of some sort is of course an open empirical question.

Interest in nanobacteria, which closely resemble nanobes, has not declined as rapidly, in part because they exhibit even more life-like characteristics. Nanobacteria are tiny, self-replicating, bacteria-shaped objects, which appear as spheres or rods when viewed under a scanning electron microscope. Initially proposed by Kajander and Çiftçioglu (1998) as explanations for certain types of pathological calcifications in human organs, nanobacteria have since been reported in wastewater, sandstones, and the stratosphere (Urbano and Urbano 2007). Nanobacteria not only react to DNA stains, strongly suggesting that they contain DNA, they also stain gram-negative, which is cited by proponents as evidence that they have cell walls. Attempts to sequence their DNA have not been very successful, however. Most importantly for skeptics, nanobacteria are extremely small, in the range 20–500 nm, too small, according to most microbiologists, to contain the biological machinery required for life.

As the discussion above indicates, the reasons cited by most microbiologists for rejecting the possibility that nanobes and nanobacteria might represent heretofore unknown forms of Earth life are (1) an inability to sequence their DNA and (2) their extremely small size. Yet as discussed in Section 9.3.3, one would not expect to be able to sequence DNA differing even modestly from that of familiar life in the identity of its nucleobases. The alleged drawback of their small size is especially intriguing, however, when one reflects that giant viruses were initially classified as bacteria on the grounds that they were too large to be viruses. Even more intriguingly, both nanobacteria and nanobes are similar in size to the mysterious carbonate globules found in ALH84001 (the famous Martian meteorite found in the Antarctic), which some astrobiologists contended, on the basis of their small size, could be the fossilized remains of ancient Martian microbes (see Section 8.3). As discussed in Sections 8.3–8.4, however, there are additional reasons for thinking that the associated carbon compounds have an abiotic origin. Seeking to settle the simmering controversies over whether nanobacteria, nanobes, and the carbonate globules in ALH84001 are too small to be living things, the National Academy of

Sciences convened a workshop to "define" the minimal size of a cell-like structure that is compatible with life. Participants concurred that the minimum size of a free-living cell is 200–300 nm (de Duve and Osborn 1999), effectively ruling out merely by stipulation nanobacteria, nanobes, and the carbonate globules in ALH84001 as candidates for unicellular organisms. What is important for our purposes, however, is that they reached this conclusion by extrapolating from the genetic and metabolic machinery found in bacteria and archaea. In short, they assumed that all free-living cells share the same molecular architecture.

There are of course no guarantees that desert varnish, nanobes, or nanobacteria represent an alternative form of Earth life. As Section 8.4 stresses, most puzzling empirical findings do not give rise to revolutionary scientific discoveries, such as the discovery of archaea and giant viruses. On the other hand, as the latter discoveries also illustrate, scientific breakthroughs often begin with empirical findings that frustrate attempts at explanation in terms of accepted entities and processes. The burden of this section has been to show that desert varnish, nanobes, and nanobacteria are genuinely anomalous in this sense: They both resemble and fail to resemble familiar life in provocative and unanticipated ways. It is thus a mistake to dismiss the *possibility* that they represent novel forms of life on the grounds that they fail to closely enough resemble familiar unicellular microbes. For the latter is just what one would expect of novel microbes descended from an alternative origin of Earth life.

9.5 Concluding Thoughts

The purpose of this chapter is not to argue that a shadow biosphere exists. It is to cast doubt on the widely accepted, rarely questioned, assumption that all life on Earth descends from a common ancestor. As Section 9.2 argued, the conjecture that Earth might be host to microbial life differing modestly from familiar life at the molecular level in the composition of some of its basic biomolecules is compatible with our current understanding of the chemical composition and molecular architecture of life. It is also compatible with prevailing models of the origin(s) of life, namely, the RNA World and the SM World. Many biologists concur that the earliest life on Earth may have exhibited significant molecular variability but nevertheless contend that (i) HGT would have amalgamated the earliest protocells into a single type of life or (ii) our form of life would have long ago eliminated any alternative forms of life in a Darwinian competition for vital resources. As we have seen, however, these arguments do not hold up well under close scrutiny. It is highly unlikely that microbes descended from alternative biogeneses would be able to exchange genes. Moreover argument (ii) fails to take into consideration what microbiologists have learned in recent years about the composition and dynamics of microbial communities. For as Section 6.3 discussed, microbes interact in a far more collaborative fashion than portrayed in classic Darwinian evolution, which takes complex multicellular eukaryotes, especially animals, as archetypical. Of all the arguments that have been advanced against the possibility of a shadow biosphere, however, the one that is viewed as most convincing is that we would have discovered such organisms by now if they exist. Yet, as Section 9.3 explained, this argument is no less problematic than the former. The tools used by microbiologists to explore the microbial world are either too crude (microscopy) or too specific (cultivation and metagenomic techniques) to distinguish a modestly different form of microbial life from familiar microbes. The upshot is that the question of whether our home planet is host to a truly novel form of microbial life, descended from an alternative biogenesis, remains an open empirical question, one well worth pursuing scientifically given the significance that finding such organisms would represent for answering the age-old question 'what is life?'

The advantages of a dedicated scientific search for shadow microbes go beyond the exciting possibility of finding an alternative form of Earth life. It is much easier to do follow-up research on a potentially biological anomaly found on Earth than one found on Mars, Europa, Titan, or Enceladus. Discovering that some or all of the potentially biological anomalies canvassed in Section 9.4 are products of as yet poorly understood abiotic processes would extend our knowledge of the ways in which chemistry and physics play out in the messy, uncontrollable world of nature. Such information would be valuable in a search for extraterrestrial life, allowing astrobiologists to recognize cases that seem potentially biological but are not. Moreover, in line with NASA's current enthusiasm for investigating Mars analog environments, a dedicated search for a shadow biosphere would help astrobiologists design and test suites of tentative criteria for hunting for potentially biological anomalies in a wide variety of extraterrestrial environments. The point is even an unsuccessful search for a shadow biosphere would make a significant contribution to the search for a second example of life.

Conclusion

This book presents a long, multifaceted argument for pursuing universal biology in the face of (in William James's colorful words) the "blooming buzzing confusion" offered by familiar Earth life to researchers. As Chapter 5 discusses, the central challenge for the program of universal biology is that familiar Earth life – the only form of life of which we can be certain – represents a single example and there are positive reasons for worrying that this example is unrepresentative of life. Biologists have discovered that life as we know it on Earth descends from a last universal common ancestor, and hence represents a single example. Moreover, biochemists have established that life elsewhere could differ from familiar life in certain ways at the molecular and biochemical levels, and they do not know how different it could be from familiar Earth life. Finally, as Chapter 6 explains, contemporary biological theorizing about life is founded upon what we now know is an unrepresentative form of familiar Earth life, namely, highly specialized, latecomers to our planet (complex multicellular eukaryotes). Indeed, a central theme of this book (Chapter 1) is that much of contemporary biological thought is still implicitly wedded to a defective, neo-Aristotelean, theoretical framework for life based on animals and plants.

As Chapter 6 discusses, what biologists are learning about the microbial world is challenging core concepts (ranging from the concept of species and the associated model of a tree of life, to the nature of the living individual and its presumed dependence upon a neo-Aristotelian conception of biological autonomy) underlying contemporary biological thought about life. Yet instead of searching for new and more fruitful theoretical concepts for generalizing about life, some biologists and many philosophers of biology cling to these older concepts, preferring to reinterpret them, sometimes weakening them almost beyond recognition, and in many cases theorizing them in an overlapping and sometimes disjunctive, pluralist fashion. In short, they embrace the disunity of life.

As Chapter 4 argues, in the context of a diversity of case studies from the history of science, the manifestation of disunity is more often a sign of a defective theoretical framework (of basic concepts for formulating general principles about a domain of phenomena) than it is a sign that no universal theory of the phenomena is possible. The danger of taking a pluralist stance towards the phenomena of life is that one will fail to recognize unity when one encounters it. For nature does not come pre-structured into theoretical categories supporting scientifically fruitful general principles. Discovering unity in the diversity of a domain of natural phenomena requires sustained theoretical and empirical research. On a pluralist stance towards life, which tends to revel in disunity, there is little incentive to do this hard work. On the other hand, it is much easier to recognize disunity (as manifested in difficult to resolve counterexamples to proposed generalizations) under an assumption of unity, than it is to recognize unity under an assumption of disunity. It is for this reason that I recommend, in Section 4.5, a monist stance towards the phenomena of biology.

There is, of course, no guarantee that the scientific program of universal biology will be successful. The monist stance is open to discovering that life is unlikely to be a unified phenomenon. It maintains, however, that the best way of making progress on this question, given our exceedingly limited experience with the phenomenon of life, is to propose and evaluate alternative theoretical frameworks for biology. To do this we need examples of life as we don't know it. As Chapter 6 argues, to the extent that it is challenging some of our most venerable, eukaryote-centric, biological concepts and principles, the microbial world is presenting us with grist for the theoretical mill. What is really needed, however, is a form of life descended from an alternative biogenesis.

The last three chapters of this book canvass three, oft discussed, possibilities for finding a truly novel form of life: (1) constructing or synthesizing life artificially (Chapter 7), (2) discovering life elsewhere in the universe, such as on a planet or moon in our solar system (Chapter 8), or (3) discovering life right here on Earth in the form of a shadow biosphere (Chapter 9). As Chapter 7 points out, any form of artificial life that we synthesize in a laboratory will be too closely based on our current Earth-centric concept of life to tell us much about the possibilities for truly novel forms of life. Moreover, the constructions of soft and hard ALife researchers are so removed from natural life that it is not at all clear that they qualify as authentically living entities. To claim that they do, on the basis of abstract resemblances between them and familiar life, is to assume that life is not a physicochemical phenomenon and yet this is the very question at issue. For as Section 4.3 discusses, the central problem facing the construction of a truly general scientific theory is identifying the right ("Goldilocks") level of abstraction for generalization, which is difficult to do based on a single example of a phenomenon.

The best options are strategies (2) and (3), that is, searching for examples of natural life as we don't know it. In this context, it strikes me that the likelihood of finding microbes on a planet or moon in our solar system are lower than that of finding a shadow biosphere. Why? As Chapter 9 discusses, we know that conditions on the early Earth actually gave rise to life, and our current understanding of life suggests that the early Earth contained at least modestly different chemical building blocks for life. On the other hand, we are not nearly as confident about whether conditions on Mars, Titan, Europa, or Enceladus could have given rise to life (for the simple reason that we do not know whether they in fact gave rise to life) and, most importantly, while some of their chemical and physical conditions resemble those on the early Earth, others are quite different. If I were to estimate the odds of the best place to look for a novel form of microbial life, descended from an alternative biogenesis, I would bet on Earth: I would search for a shadow biosphere.

References

Aagaard, K., J. Ma, K. M. Antony, R. Ganu, J. Petrosino, and J. Versalovic, The placenta harbors a unique microbiome, *Sci. Transl. Med.*, **6**, 237ra65, 2014.

Agmon, E., A. J. Gates, V. Churavy, and R. Beer, Exploring the space of viable configurations in a model of metabolism–boundary co-construction, *Artif. Life*, **22**, 153–171, 2016.

Ainsworth, G. C., *Introduction to the History of Mycology*, Cambridge University Press, Cambridge, 1976.

Airapetian, V. S., A. Glocer, G. V. Khazanov, et al., How hospitable are space weather affected habitable zones? The role of ion escape, *Astrophys. J.*, **836**, L3, 2017.

Alberts, B., A. Johnson, J. Lewis, M. Raff, K. Roberts, and P. Walter, *Molecular Biology of the Cell*, 4th edition, Garland Science, New York, 2002.

Allwood, A. C., J. P. Grotzinger, A. H. Knoll, I. W. Burch, and M. S. Anderson, Controls on development and diversity of early Archean stromatolites, *Proc. Natl. Acad. Sci. USA*, **106**, 9548–9555, 2009.

Amend, J. P., and E. L. Shock, Energetics of amino acid synthesis in hydrothermal ecosystems, *Science*, **281**, 1659–1662, 1998.

American Society for Microbiology, Humans have ten times more bacteria than human cells: How do microbial communities affect human health? *ScienceDaily*, **5**, 2008. (www.sciencedaily.com/releases/2008/06/080603085914.htm)

Anastasi, C., F. F. Buchet, M. A. Crowe, A. L. Parkes, M. W. Powner, J. M. Smith, and J. D. Sutherland, RNA: prebiotic product, or biotic invention? *Chem. Biodivers.*, **4**, 721–739, 2007.

Anbar, A. D., K. J. Zahnle, G. L. Arnold, and S. J. Mojzsis, Extraterrestrial iridium, sediment accumulation and the habitability of the early Earth's surface, *J. Geophys. Res.*, **106**, 3219–3236, 2001.

Ander, M. E., M. A. Zumberge, T. Lautzenhiser, R. L. Parker, C. L. V. Aiken, M. R. Gorman, M. M. Nieto, A. P. R. Cooper, and W. Wirtz, Test of Newton's inverse-square law in the Greenland ice cap, *Phys. Rev. Lett.*, **62**, 985–988, 1989.

Andersen, A., and H. Haack, Carbonaceous chondrites: tracers of the prebiotic chemical evolution of the Solar System, *Int. J. Astrobiol.*, **4**, 13–17, 2005.

Anderson, C., The end of theory: the data deluge makes the scientific method obsolete, *Wired Magazine*, July 16, 2008. (www.wired.com/science/discoveries/magazine/16 07/pb theory)

Arslan, D., M. Legendre, V. Seltzer, C. Abergel, and J. M. Claverie, Distant Mimivirus relative with a larger genome highlights the fundamental features of Megaviridae, *Proc. Natl. Acad. Sci. USA*, **108**, 17486–17491, 2011.

Atwood, K. C., L. K. Schneider, and F. J. Ryan, Selective mechanisms in bacteria, *Cold Spring Harb. Symp. Quant. Biol.*, **16**, 345–354, 1951.

Audi, R., ed., *The Cambridge Dictionary of Philosophy*, Cambridge University Press, New York, 1995.

Bada, J., State-of-the-art instruments for detecting extraterrestrial life, *Proc. Natl. Acad. Sci. USA*, **98**, 797–800, 2001.

Bains, W., Many chemistries could be used to build living systems, *Astrobiology*, **4**, 137–167, 2004.

Baker, B. J., G. W. Tyson, R. I. Webb, J. Flanagan, P. Hugenholtz, E. E. Allen, and J. F. Banfield, Lineages of acidophilic Archaea revealed by community genomic analysis, *Science*, **314**, 1933–1935, 2006.

Ball, P., *H₂O: A Biography of Water*, Weidenfeld & Nicolson, London, 1999.

Bamford, D. H., R. M. Burnett, and D. I. Stuart, Evolution of viral structure, *Theor. Popul. Biol.*, **61**, 461–470, 2002.

Barnes, J., ed., *The Complete Works of Aristotle*, Princeton University Press, Princeton, NJ, 1984.

Baross, J. A., S. A. Benner, G. D. Cody, et al., *The Limits to Organic Life in Planetary Systems*, National Academy Press, Washington, DC, 2007.

Beatty, J., The evolutionary contingency thesis, in *Concepts, Theories, and Rationality in the Biological Sciences: The Second Pittsburgh-Konstanz Colloquium in the Philosophy of Science*, pp. 45–81, G. Wolters and J. Lennox, eds., University of Pittsburgh Press, Pittsburgh, PA, 1995.

Bedau, M. A., Weak emergence, *Philos. Perspect.*, **11**, 375–399, 1997.

Bedau, M. A., Four puzzles about life, *Artif. Life*, **4**, 125–140, 1998. [Reprinted in Bedau and Cleland 2010]

Bedau, M. A., An Aristotelian account of minimal life, *Astrobiology*, **10**, 1011–1020, 2010.

Bedau, M. A., and C. E. Cleland, eds., *The Nature of Life: Classic and Contemporary Perspectives from Philosophy and Science*, Cambridge University Press, Cambridge, 2010.

Bedau, M. A., J. S. McCaskill, N. H. Packard, et al., Open problems in artificial life, *Artif. Life*, **6**, 363–376, 2000.

Beegle, L. W., M. G. Wilson, F. Abilleira, J. F. Jordan, and G. R. Wilson, A concept for NASA's 2016 astrobiology field laboratory, *Astrobiology*, **7**, 545–577, 2007.

Bell, E. A., P. Boehnke, T. M. Harrison, and W. L. Mao, Potentially biogenic carbon preserved in a 4.1 billion-year-old zircon, *Proc. Natl. Acad. Sci. USA*, **112**, 14518–14521, 2015.

Bell, M. S., Experimental shock decomposition of siderite and the origin of magnetite in Martian meteorite ALH 84001, *Meteorit. Planet. Sci.*, **42**, 935–940, 2007.

Belloche, A., R. T. Garrod, H. S. Müller, and K. M. Menten, Detection of a branched alkyl molecule in the interstellar medium: iso-propyl cyanide, *Science*, **345**, 1584–1587, 2014.

Belshaw, R., V. Pereira, A. Katzourakis, G. Talbot, J. Paces, A. Burt, and M. Tristem, Long-term reinfection of the human genome by endogenous retroviruses, *Proc. Natl. Acad. Sci. USA*, **101**, 4894–4899, 2004.

Benner, S. A., Redesigning life: organic chemistry and the evolution of protein, *Chimia*, **41**, 142–148, 1987.

Benner, S. A., Expanding the genetic lexicon: incorporating nonstandard amino acids into proteins by ribosome-based synthesis, *Trends Biotechnol.*, **12**, 158–163, 1994.

Benner, S. A., Understanding nucleic acids using synthetic chemistry, *Acc. Chem. Res.*, **37**, 784–797, 2004.

Benner, S. A., Defining life, *Astrobiology*, **10**, 1021–1030, 2010.

Benner, S. A., and A. D. Ellington, The last ribo-organism, *Nature*, **329**, 296–296, 1987.

Benner, S. A., and D. Hutter, Phosphates, DNA, and the search for nonterran life: a second generation model for genetic molecules, *Bioorg. Chem.*, **30**, 62–80, 2002.

Benner, S. A., K. G. Devine, L. N. Matveeva, and D. H. Powell, The missing organic molecules on Mars, *Proc. Natl. Acad. Sci. USA*, **97**, 2425–2430, 2000.

Benner, S. A., A. Ricardo, and M. A. Carrigan, Is there a common chemical model for life in the universe? *Curr. Opin. Chem. Biol.*, **8**, 672–689, 2004. [Reprinted in Bedau and Cleland 2010]

Bianciardi, G., J. D. Miller, P. A. Straat, and G. V. Levin, Complexity analysis of the Viking labeled release experiments, *Int. J. Aeronaut. Space*, **13**, 14–26, 2012.

Bich, L., and S. Green, Is defining life pointless? Operational definitions at the frontiers of biology, *Synthese*, **195**, 3919–3946, 2017.

Biemann, K., J. Oro, P. Toulmin III, L. E. Orgel, A. O. Nier, D. M. Anderson, P. G. Simmonds, D. Flory, A. V. Diaz, D. R. Rushneck, and J. A. Biller, Search for organic and volatile inorganic compounds in two surface samples from the Chryse Planitia region of Mars, *Science*, **194**, 72–76, 1976.

Biemann, K., J. Oro, P. Toulmin III, et al., The search for organic substances and inorganic volatile compounds in the surface of Mars, *J. Geophys. Res.*, **82**, 4641–4658, 1977.

Birtles, R. J., T. J. Rowbotham, C. Storey, T. J. Marrie, and D. Raoult, Chlamydia-like obligate parasite of free-living amoebae, *Lancet*, **349**, 925–926, 1997.

Boden, M., *The Philosophy of Artificial Life*, Oxford University Press, New York, 1996.

Boden, M., Autopoiesis and life, *Cogn. Sci. Q.*, **1**, 117–145, 2000.

Boto, L., Horizontal gene transfer in evolution: facts and challenges, *Proc. R. Soc. B*, **277**, 819–827, 2010.

Boto, L., Horizontal gene transfer in the acquisition of novel traits by metazoans, *Proc. R. Soc. B*, **281**, 2013.2450, 2014.

Boyd, R., Kinds, complexity and multiple realization, *Philos. Stud.*, **95**, 67–98, 1999.

Bridgman, P. W., Operational analysis, *Philos. Sci.*, **5**, 114–131, 1938.

Broad, C. D., *The Mind and its Place in Nature*, Routledge and Kegan Paul, London, 1925.

Brock, T. D., *Thermophilic Microorganisms and Life at High Temperatures*, Springer Verlag, Berlin, 1978.

Budin, I., and N. K. Devaraj, Membrane assembly driven by a biomimetic coupling reaction, *J. Am. Chem. Soc.*, **134**, 751–753, 2012.

Bull, M., and N. T. Plummer, Part 1: the human gut microbiome in health and disease, *Integr. Med. (Encinitas)*, **13**, 17–22, 2014.

Burke, C., P. Steinberg, D. Rusch, S. Kjelleberg, and T. Thomas, Bacterial community assembly based on functional genes rather than species, *Proc. Natl. Acad. Sci. USA*, **108**, 14288–14293, 2011.

Cady, S., and N. Noffke, Geobiology: evidence for early life on Earth and the search for life on other planets, *GSA Today*, **19**, 4–10, 2009.

Cafferty, B. J., and N. V. Hud, Abiotic synthesis of RNA in water: a common goal of prebiotic chemistry and bottom-up synthetic biology, *Curr. Opin. Chem. Biol.*, **22**, 146–157, 2014.

Cairns-Smith, A., *Genetic Takeover and the Mineral Origin of Life*, Cambridge University Press, Cambridge, 1982.

Cairns-Smith, A. G., A. J. Hall, and M. J. Russell, Mineral theories of the origin of life and an iron sulfide example, *Orig. Life Evol. Biosph.*, **22**, 161–180, 1992.

Callahan, M. P., K. E. Smith, J. Cleaves II, et al., Carbonaceous meteorites contain a wide range of extraterrestrial nucleobases, *Proc. Natl. Acad. Sci. USA*, **108**, 13995–13998, 2011.

Canfield, D. E., The early history of atmospheric oxygen, *Annu. Rev. Earth Planet. Sci.*, **33**, 1–36, 2005.

Carnegie Institution for Science, *Re-conceptualizing the Origins of Life, Workshop*, November 9–13, 2015. (https://carnegiescience.edu/events/lectures/re-conceptualiz ing-origin-life)

Cartwright, N., *How the Laws of Physics Lie*, Oxford University Press, Oxford, 1983.

Cartwright, N., *Nature's Capacities and Their Measurements*, Oxford University Press, Oxford, 1989.

Caschera, F., M. A. Bedau, A. Buchanan, et al., Coping with complexity: machine learning optimization of cell-free protein synthesis, *Biotechnol. Bioeng.*, **108**, 2218–2228, 2011.

Cash, H. L., C. V. Whitman, C. L. Behrendt, and L. V. Hooper, Symbiotic bacteria direct expression of an intestinal bactericidal lectin, *Science*, **313**, 1126–1130, 2006.

Cech, T. R., Self-splicing and enzymatic activity of an intervening sequence RNA from *Tetrahymena*, *Nobel Lecture*, Nobel Media, 1989. (www.nobelprize.org/prizes/chem istry/1989/cech/lecture)

Cech, T. R., The RNA worlds in context, *Cold Spring Harb. Perspect. Biol.*, **4**, a006742, 2012.

Chalmers, A., Galilean relativity and Galileo's relativity, in *Correspondence, Invariance and Heuristics: Essays in Honor of Heinz Post*, pp. 189–205, S. French and J. Kamminga, eds., Springer, Dordrecht, 1993.

Chalmers, D. J., and F. Jackson, Conceptual analysis and reductive explanation, *Philos. Rev.*, **110**, 315–360, 2001.

Chang, H., *Is Water H₂O? Evidence, Realism, and Pluralism*, Springer, Dordrecht, 2014.

Chao, L., The meaning of life, *BioScience*, **50**, 245–250, 2000.

Chatterjee, S., A symbiotic view of the origin of life at hydrothermal impact crater-lakes, *Phys. Chem. Chem. Phys.*, **18**, 20033–20046, 2016.

Chicote, E., A. M. García, D. A. Moreno, M. I. Sarró, P. I. Lorenzo, and F. Montero, Isolation and identification of bacteria from spent nuclear fuel pools, *J. Ind. Micro-biol. Biotechnol.*, **32**, 155–162, 2005.

Chyba, C. F., and G. D. McDonald, The origin of life in the solar system: current issues, *Annu. Rev. Earth Planet. Sci.*, **23**, 215–249, 1995.

Clark, E., The problem of biological individuality, *Biol. Theor.*, **5**, 312–325, 2010.

Claus, D., *Toward the Soul: An Inquiry into the Meaning of Ψυχή Before Plato*, Yale University Press, New Haven, CT, 1981.

Cleland, C. E., Space: an abstract system of non-supervenient relations, *Philos. Stud.*, **46**, 19–40, 1984.

Cleland, C. E., Is the Church–Turing thesis true?, *Minds Mach.*, **3**, 283–312, 1993.

Cleland, C. E., The concept of computability, *Theor. Comput. Sci.*, **317**, 209–225, 2004.

Cleland, C. E., Understanding the nature of life: a matter of definition or theory?, in *Life As We Know It*, pp. 589–600, J. Seckbach, ed., Springer, Dordrecht, 2006.

Cleland C. E., Epistemological issues in the study of microbial life: alternative terran biospheres?, *Stud. Hist. Philos. Biol. Biomed. Sci.*, **38**, 847–861, 2007.

Cleland, C. E., Life without definitions, *Synthese*, **185**, 125–144, 2012.

Cleland, C. E., Conceptual challenges for contemporary theories of the origin of life, *Curr. Org. Chem.*, **17**, 1704–1709, 2013.

Cleland, C. E., Moving beyond definitions in the search for extraterrestrial life, *Astrobiology*, **19**, 722–729, 2019.

Cleland, C. E., and C. F. Chyba, Defining 'life', *Orig. Life Evol. Biosph.*, **32**, 397–393, 2002.

Cleland, C. E., and C. F. Chyba, Does 'life' have a definition?, in *Planets and Life: The Emerging Science of Astrobiology*, pp. 119–131, W. T. Sullivan III and J. A. Baross, eds., Cambridge University Press, Cambridge, 2007. [Reprinted in Bedau and Cleland 2010]

Cleland C. E., and S. D. Copley, The possibility of alternative microbial life on earth, *Int. J. Astrobiol.*, **2**, 165–173, 2005. [Reprinted in Bedau and Cleland 2010]

Cody, G. D., Transition metal sulfides and the origin of metabolism, *Annu. Rev. Earth Planet. Sci.*, **32**, 569–599, 2004.

Cohen, I. B., Newton's concepts of force and mass, with notes on the laws of motion, in *The Cambridge Companion to Newton*, pp. 57–84, I. B. Cohen and G. E. Smith, eds., Cambridge University Press, Cambridge, 2002.

Cohen, I. B., and G. E. Smith, eds., *The Cambridge Companion to Newton*, Cambridge University Press, Cambridge, 2002.

Coleman, W., *Biology in the Nineteenth Century: Problems of Form, Function, and Transformation*, Cambridge University Press, Cambridge, 1977.

Conrad, P. G., and K. H. Nealson, A non-earth-centric approach to life detection, *Astrobiology*, **1**, 15–24, 2001.

Copley, S. D., E. Smith, and H. J. Morowitz, A mechanism for the association of amino acids and their codons and the origin of the genetic code, *Proc. Natl. Acad. Sci. USA*, **102**, 4442–4447, 2005.

Copley, S. D., E. Smith, and H. J. Morowitz, The origin of the RNA world: co-evolution of genes and metabolism, *Bioorg. Chem.*, **35**, 430–443, 2007.

Cornish-Bowden, A., G. Piedrafita, M. Morán, M. L Cárdenas, and F. Montero, Simulating a model of metabolic closure, *Biol. Theor.*, **8**, 383–390, 2013.

Costello, E. K., K. Stagaman, L. Dethlefsen, B. J. M. Bohannan, and D. A. Relman, The application of ecological theory towards an understanding of the human microbiome, *Science*, **336**, 1255–1262, 2012.

Craver, C. F., Beyond reduction: mechanisms, multifield integration and the unity of neuroscience, *Stud. Hist. Philos. Biol. Biomed. Sci.*, **36**, 373–395, 2005.

Crick, F., On protein synthesis, *Symp. Soc. Exp. Biol.*, **12**, 138–167, 1958.

Crick, F. H. C., The origin of the genetic code, *J. Mol. Biol.*, **38**, 367–379, 1968.

Crick, F. H. C., *Life Itself*, Simon and Schuster, New York, 1981.

Cronin, J. R., and S. Pizzarello, Amino acids in meteorites, *Adv. Space Res.*, **3**, 5–18, 1983.

Da Costa, N. C. A., and S. French, *Science and Partial Truth*, Oxford University Press, Oxford, 2003.

Dagan, T., M. Roettger, D. Bryant, and W. Martin, Genome networks root the tree of life between prokaryotic domains, *Genome Biol. Evol.*, **12**, 379–392, 2010.

Damiano, L., and P. Luisi, Towards an autopoietic redefinition of life, *Orig. Life Evol. Biosph.*, **40**, 145–149, 2010.

Dang, H., and C. R. Lovell, Microbial surface colonization and biofilm development in marine environments, *Microbiol. Mol. Biol. Rev.*, **80**, 91–138, 2015.

Darwin, C., *On the Origin of Species by Means of Natural Selection, or Preservation of Favoured Races in the Struggle for Life*, John Murray, London, 1859.

Darwin, C., *The Descent of Man*, John Murray, London, 1871.

Davey, M. E., and G. A. O'toole, Microbial biofilms: from ecology to molecular genetics, *Microbiol. Mol. Biol. Rev.*, **64**, 847–867, 2000.

Davies, P. C. W., *The Eerie Silence*, Houghton Mifflin Harcourt, New York, 2010.

Davies, P. C. W., and C. H. Lineweaver, Finding a second sample of life on earth, *Astrobiology*, **5**, 154–163, 2005.

Davies, P. C. W., S. A. Benner, C. E. Cleland, C. H. Lineweaver, C. P. McKay, and F. Wolfe-Simon, Signatures of a shadow biosphere, *Astrobiology*, **9**, 241–249, 2009.

Dawkins, R., *The Selfish Gene*, Oxford University Press, Oxford, 1976.

Dawkins, R., Universal Darwinism, in *Evolution from Molecules to Men*, pp. 403–425, D. S. Bendall, ed., Cambridge University Press, Cambridge, 1983. [Reprinted in Bedau and Cleland 2010]

Deamer, E. W., and J. P. Dworkin, Chemistry and physics of primitive membranes, *Top. Curr. Chem.*, **259**, 1–27, 2005.

Decho, A. W., and T. Gutierrez, Microbial extracellular polymeric substances (EPSs) in ocean systems, *Front. Microbiol.*, **8**, 922, 2017.

de Duve, C., *Vital Dust: The Origin and Evolution of Life*, Basic Books, New York, 1995.

de Duve C., and M. J. Osborn, Panel 1: discussion, in *Size Limits of Very Small Microorganisms: Proceedings of a Workshop*, National Academies Press, Washington, DC, 1999.

Dennett, D. C., *Darwin's Dangerous Idea: Evolution and the Meanings of Life*, Simon and Schuster, New York, 1995.

Dobson, C. M., G. B. Ellison, A. F. Tuck, and V. Vaida, Atmospheric aerosols as prebiotic chemical reactors, *Proc. Natl. Acad. Sci. USA*, **97**, 11864–11868, 2000.

Donaldson, D. J., H. Tervahattu, A. F. Tuck, and V. Vaida, Organic aerosols and the origin of life: an hypothesis, *Orig. Life Evol. Biosph.*, **34**, 57–67, 2004.

Doolittle, W. F., Phylogenetic classification and the universal tree, *Science*, **284**, 2124–2128, 1999.

Doolittle, W. F., Darwinizing Gaia, *J. Theor. Biol.*, **434**, 11–19, 2017.

Doolittle, W. F., and A. Booth, It's the song, not the singer: an exploration of holobionts and evolutionary theory, *Biol. Philos.*, **32**, 5–24, 2017.

Doolittle, W. F., and R. T. Papke, Genomics and the bacterial species problem, *Genome Biol.*, **7**, 116, 2006.

Douglas, A. E., *Fundamentals of Microbiome Science*, Princeton University Press, Princeton, NJ, 2018.

Downes, S., The importance of models in theorizing: a deflationary semantic view, in *Proceedings of the 1992 Biennial Meeting of the Philosophy of Science Association*, volume 1, pp. 142–153, D. Hull, M. Forbes, and K. Okruhlik, eds., University of Chicago Press, Chicago, IL, 1992.

Dupré, J., *The Disorder of Things*, Harvard University Press, Cambridge, MA, 1993.

Dupré, J., *The Selfish Gene Meets the Friendly Germ*, Keynote speech at the 22nd Regional Conference on the History and Philosophy of Science, Boulder, CO, 2007.

Dupré, J., *Processes of Life*, Oxford University Press, Oxford, 2012.

Dupré, J., and M. O'Malley, Metagenomics and biological ontology, *Stud. Hist. Philos. Biol. Biomed. Sci.*, **38**, 834–846, 2007a.

Dupré, J., and M. O'Malley, Size doesn't matter: towards a more inclusive philosophy of biology, *Biol. Philos.*, **22**, 155–191, 2007b.

Dupré, J., and M. O'Malley, Varieties of living things: life at the intersection of lineage and metabolism, *Philos. Theor. Biol.*, **1**, e003, 2009.

Dupressoir, A., C. Lavialle, and T. Heidmann, From ancestral infectious retroviruses to bona fide cellular genes: role of the captured syncytins in placentation, *Placenta*, **33**, 663–671, 2012.

Dyson, F. J., *Origins of Life*, Cambridge University Press, Cambridge, 1999.

Earman, J., J. T. Roberts, and S. Smith, Ceteris paribus lost, *Erkenntnis*, **57**, 281–301, 2002.

Eckhardt, D. H., C. Jekeli, A. R. Lazarewicz, A. J. Romaides, and R. W. Sands, Experimental evidence for a violation of Newton's inverse-square laws of gravitation, *Eos*, **69**, 1046, 1988.

Ehrenfreund, P., and J. Cami, Cosmic carbon chemistry: from the interstellar medium to the early earth, *Cold Spring Harb. Perspect. Biol.*, **2**, a002097, 2010.

Ehrenfreund, P., and K. M. Menten, From molecular clouds to the origin of life, in *Astrobiology: The Quest for the Conditions of Life*, pp. 7–23, G. Horneck and C. Baumstark-Khan, eds., Springer, Berlin, 2002.

Eigen, M., *Steps Towards Life: A Perspective on Evolution*, Oxford University Press, Oxford, 1992.

Eigenbrode, J. L., R. E. Summons, A. Steele, et al., Organic matter preserved in 3-billion-year-old mudstones at Gale crater, Mars, *Science*, **360**, 1096–1101, 2018. (And supplementary materials)

El Albani, A., S. Bengrson, D. E. Canfield, et al., Large colonial organisms with coordinated growth in oxygenated environments 2.1 Gyr ago, *Nature*, **466**, 100–104, 2010.

Ellison, C. K., T. N. Dalia, A. V. Ceballos, J. C.-Y. Wang, N. Biais, Y. V. Brun, and A. B. Dalia, Retraction of DNA-bound type IV competence pili initiates DNA uptake during natural transformation in *Vibrio cholerae*, *Nature Microbiol.*, **3**, 773–780, 2018.

El-Naggar, M. Y., and S. E. Finkel, Live wires, *The Scientist*, **27**, 38–43, 2013.

El-Naggar, M. Y., G. Wanger, K. M. Leung, T. D. Yuzvinsky, G. Southam, J. Yang, W. M. Lau, N. K. H. Nealson, and Y. A. Gorby, Electrical transport along bacterial nanowires from *Schewanella oneidensis* MR-1, *Proc. Natl. Acad. Sci. USA*, **107**, 18127–18131, 2010.

England, J. L., Statistical physics of self-replication, *J. Chem. Phys.*, **139**, 121923, 2013.

Ereshefsky, M., *The Poverty of the Linnaean Hierarchy: A Philosophical Study of Biological Taxonomy*, Cambridge University Press, Cambridge, 2001.

Ereshefsky, M., Microbiology and the species problem, *Biol. Philos.*, **25**, 553–568, 2010.

Ereshefsky, M., and M. Pedroso, Rethinking evolutionary individuality, *Proc. Natl. Acad. Sci. USA*, **112**, 10126–10132, 2015.

Ereshefsky, M., and M. Pedroso, What biofilms can teach us about individuality, in *Individuals Across the Sciences*, pp. 85–102, A. Guay and T. Pradeu, eds., Oxford University Press, Oxford, 2016.

Faivre, D., and D. Schüler, Magnetotactic bacteria and magnetosomes, *Chem. Rev.*, **108**, 4875–4898, 2008.

Falcon, A., *Aristotle and the Science of Nature: Unity Without Uniformity*, Cambridge University Press, Cambridge, 2005.

Feinberg, G., and R. Shapiro, *Life Beyond Earth: Intelligent Earthlings, Guide to the Universe*, William Morrow, New York, 1980.

Ferris, J. P., Montmorillonite-catalysed formation of RNA oligomers: the possible role of catalysis on the origins of life, *Philos. Trans. R. Soc. B*, **361**, 1777–1786, 2006.

Flemming, H. C., and J. Wingender, The biofilm matrix, *Nature Rev. Microbiol.*, **8**, 623–633, 2010.

Forterre, P., To be or not to be alive: how recent discoveries challenge the traditional definitions of viruses and life, *Stud. Hist. Philos. Sci. C, Stud. Hist. Philos. Biol. Biomed. Sci.*, **59**, 100–108, 2016.

Foster, J. A., and K. M. Neufeld, Gut-brain axis: how the microbiome influences anxiety and depression, *Trends Neurosci.*, **36**, 305–312, 2013.

Franklin, L. R., Bacteria, sex, and systematics, *Philos. Sci.*, **74**, 69–95, 2007.

Frigg, R., Scientific representation and the semantic view of theories, *Theoria*, **55**, 49–65, 2006.

Frigg, R., and S. Hartmann, Models in science, in *The Stanford Encyclopedia of Philosophy*, E. N. Zalta, ed., 2012. (https://plato.stanford.edu/archives/sum2018/entries/models-science)

Fry, I., Are the different hypotheses on the emergence of life as different as they seem?, *Biol. Philos.*, **10**, 389–417, 1995. [Reprinted in Bedau and Cleland 2010]

Fry, I., *The Emergence of Life on Earth*, Rutgers University Press, New Brunswick, NJ, 2000.

Furley, D., *Self-Motion: From Aristotle to Newton*, Princeton University Press, Princeton, NJ, 1994.

Galhardo, R. S., P. J. Hastings, and S. M. Rosenberg, Mutations as a stress response and the regulation of evolvability, *Crit. Rev. Biochem. Mol. Biol.*, **42**, 399–435, 2007.

Gánti, T., *The Principle of Life*, with commentary by J. Griesemer and E. Swathmáry, Oxford University Press, New York, 2003. [Selections reprinted in Bedau and Cleland 2010]

Gao, C., X. Ren, A. S. Mason, H. Liu, M. Xiao, J. Li, and D. Fu, Horizontal gene transfer in plants, *Funct. Integr. Genomics*, **14**, 23–29, 2014.

Gesteland, R. F., and J. F. Atkins, eds., *The RNA World*, Cold Spring Harbor Laboratory Press, New York, 1993.

Gibson, D. G., J. I. Glass, C. Lartigue, et al., Creation of a bacterial cell controlled by a chemically synthesized genome, *Science*, **329**, 52–56, 2010.

Giere, R., *Understanding Scientific Reasoning*, Holt, Rinehart and Winston, New York, 1979.

Giere, R., *Explaining Science: A Cognitive Approach*, University of Chicago Press, Chicago, IL, 1988.

Gilbert, W., The RNA World, *Nature*, **319**, 618, 1986.

Gilbert, S. F., J. Sapp, and A. I. Tauber, A symbiotic view of life: we have never been individuals, *Q. Rev. Biol.*, **87**, 325–341, 2012.

Glasser, N. R., S. H. Saunders, and D. K. Newman, The colorful world of extracellular electron shuttles, *Annu. Rev. Microbiol.*, **71**, 731–751, 2017.

Glavin, D. P., M. Schubert, O. Botta, G. Kminek, and J. Bada, Detecting pyrolysis products from bacteria on Mars, *Earth Planet. Sci. Lett.*, **185**, 1–5, 2001.

Glavin, D. P., J. P. Dworkin, A. Aubrey, et al., Amino acid analyses of Antarctic CM2 meteorites using liquid chromatography-time of flight-mass spectrometry, *Meteorit. Planet. Sci.*, **41**, 889–902, 2006.

Godfrey-Smith, P., *Darwinian Populations and Natural Selection*, Oxford University Press, Oxford, 2009.

Godfrey-Smith, P., Darwinian individuals, in *From Groups to Individuals: Evolution and Emerging Individuality*, pp. 17–36, F. Bouchard and P. Huneman, eds., MIT Press, Cambridge, MA, 2013.

Golden, D. C., D. W. Ming, C. S. Schwandt, H. V. Lauer, R. A. Socki, R. V. Morris, G. E. Lofgren, and G. A. McKay, A simple inorganic process for formation of carbonates, magnetite, and sulfides in Martian meteorite ALH84001, *Am. Mineral.*, **8**, 370–375, 2001.

Goldenfeld, N., and C. Woese, Life is physics: evolution as a collective phenomenon far from equilibrium, *Annu. Rev. Condens. Matter Phys.*, **2**, 375–399, 2011.

Gould, S. J., *Wonderful Life: The Burgess Shale and the Nature of History*, W.W. Norton, New York, 1990.

Grene, M., and D. Depew, *The Philosophy of Biology: An Episodic History*, Cambridge University Press, Cambridge, 2004.

Griesemer, J. R., Modeling in the museum: on the role of remnant models in the work of Joseph Grinnell, *Biol. Philos.*, **5**, 3–36, 1990.

Griesemer, J. R., The units of evolutionary transition, *Selection*, **1**, 67–80, 2000.

Grinspoon, D., *Venus Revealed: A New Look Below the Clouds of Our Mysterious Twin Planet*, Perseus, Cambridge, 1997.

Grinspoon, D., *Lonely Planets: The Natural Philosophy of Alien Life*, HarperCollins, New York, 2003.

Groff, R., and J. Greco, eds., *Power and Capacities in Philosophy: The New Aristotelianism*, Routledge, New York, 2013.

Guthrie, W. K. C., *A History of Greek Philosophy*, volume 6, Cambridge University Press, Cambridge, 1990.

Guyer, P., and E. Matthews, trans., *Imanuel Kant, Critique of the Teleological Power of Judgement*, Cambridge University Press, Cambridge, 2001. [Reprinted in Bedau and Cleland 2010]

Hacking, I., *Representing and Intervening: Introductory Topics in the Philosophy of Natural Science*, Cambridge University Press, Cambridge, 1983.

Hagen, J. B., Five kingdoms, more or less: Robert Whittaker and the broad classification of organisms, *BioScience*, **62**, 62–74, 2012.

Hall, T. S., trans., *René Descartes, Treatise of Man*, Prometheus Books, New York, 2003.

Hays, L., ed., *NASA Astrobiology Strategy*, NASA Astrobiology Institute, 2015.

Hazen, R. M., *Genesis: The Scientific Quest for Life's Origin*, Joseph Henry Press, Washington, DC, 2005.

Hazen, R. M., D. Papineau, W. Bleeker, et al., Mineral evolution, *Am. Mineral.*, **93**, 1693–1720, 2008.

Hecht, M. H., S. P. Kounaves, R. C. Quinn, et al., Detection of perchlorate and the soluble chemistry of Martian soil at the Phoenix lander site, *Science*, **325**, 64–67, 2009.

Hennet, R. J. C., N. G. Holm, and M. H. Engel, Abiotic synthesis of amino acids under hydrothermal conditions and the origin of life: a perpetual phenomenon?, *Naturwissenschaften*, **79**, 361–365, 1992.

Hermes, H., *Eine Axiomatisierung der Allgemeinen Mechanik*, Hirzel, Leipzig, 1938.

Hermes, H., Modal operators in an axiomatisation of mechanics, *Proceedings of the Colloque International sur la Méthode Axiomatique Classique et Moderne*, Paris, 1959.

Heward, A., Life on Titan: stand well back and hold your nose!, *PhysOrg.com*, 2010.

Hoelzer, G. A., E. Smith, and J. W. Pepper, On the logical relationship between natural selection and self organization, *J. Evol. Biol.*, **19**, 1785–1794, 2006.

Hoffman, P. F., and D. Schrag, The snowball Earth hypothesis: testing the limits of global change, *Terra Nova*, **14**, 129–155, 2002.

Holm, N.G., and E. M. Andersson, Organic molecules on the primitive Earth: hydrothermal systems, in *The Molecular Origins of Life: Assembling Pieces of the Puzzle*, pp. 86–99, A. Brack, ed., Cambridge University Press, Cambridge, 1998.

Hong, S. H., Y. C. Kwon, and M. C. Jewett, Non-standard amino acid incorporation into proteins using *Escherichia coli* cell-free protein synthesis, *Front. Chem.*, **2**, 34.10.3389, 2014.

Hugenholtz, P., B. M. Goebel, and N. R. Pace, Impact of culture-independent studies on the emerging phylogenetic view of bacterial diversity, *J. Bacteriol.*, **180**, 4765–4774, 1998.

Hull, D. L., Individuality and selection, *Annu. Rev. Ecol. Evol. Syst.*, **11**, 311–332, 1980.

Hull, D. L., On the plurality of species: questioning the party line, in *Species: New Interdisciplinary Essays*, pp. 23–48, R. Wilson, ed., MIT Press, Cambridge, MA, 1999.

Hume, D., *A Treatise on Human Nature*, L. A. Selby-Bigge, ed., Oxford University Press, Oxford, 1888.

Humphreys, P., Aspects of emergence, *Philos. Topics*, **24**, 53–70, 1996.

Hunt, K. M., J. A. Foster, L. J. Forney, U. M. E. Schütte, D. L. Beck, Z. Abdo, L. K. Fox, J. E. Williams, M. K. McGuire, and M. A. McGuire, Characterization of the diversity and temporal stability of bacterial communities in human milk, *PLoS One*, **6**, e21313, 2011.

Hunten, D. M., Possible oxidant sources in the atmosphere and surface of Mars, *J. Mol. Evol.*, **14**, 71–78, 1979.

Hutchison, C. A., R.-Y. Chuang, V. N. Noskov, et al., Design and synthesis of a minimal bacterial genome, *Science*, **351**, 1371–1494, 2016.

Huxley, J., *The Modern Evolutionary Synthesis*, George Allen and Unwin, London, 1942.

Huxley, T. H., Criticism on "The Origin of Species" [1864], in *Darwiniana*, Macmillan, London, 1893.

Hystad, G., R. T. Downs, and R. M. Hazen, Mineral species frequency distribution conforms to a large number of rare events model: prediction of earth's missing minerals, *Math. Geosci.*, **47**, 647–661, 2015.

Jablonka, E., and M. J. Lamb, *Evolution in Four Dimensions: Genetic, Epigenetic, Behavioral, and Symbolic Variation in the History of Life*, MIT Press, Cambridge, MA, 2005.

Jacob, G. F., *The Logic of Life: A History of Heredity*, Princeton University Press, Princeton, NJ, 1973.

James, W., *The Principles of Psychology*, Henry Holt, New York, 1890.

Jiang, L., P. Dziedzic, Z. Spacil, et al., Abiotic synthesis of amino acids and self-crystallization under prebiotic conditions, *Sci. Rep.*, **4**, 6769, 2014.

Johnson, S. K., M. A. Fitza, D. A. Lerner, D. M. Calhoun, M. A. Beldon, E. T. Chan, and P. T. J. Johnson, Risky business: linking *Toxoplasma gondii* infection and entrepreneurship behaviors across individuals and countries, *Proc. R. Soc. B*, **285**, 2018.0822, 2018.

Joyce, G. F., Foreword, in *Origins of Life: The Central Concepts*, pp. xi–xii, D. W. Deamer and G. R. Fleischaker, eds., Jones & Bartlett, Boston, MA, 1994.

Joyce, G. F., and L. E. Orgel, Prospects for understanding the origin of the RNA World, in *The RNA World*, 2nd edition, pp. 49–78, R. F. Gesteland, T. R. Cech, and J. F. Atkins, eds., Cold Spring Harbor Laboratory Press, New York, 1999.

Kajander, E. O., and N. Çiftçioglu, Nanobacteria: an alternative mechanism for pathogenic intra and extracellular calcification and stone formation, *Proc. Natl. Acad. Sci. USA*, **95**, 8274–8279, 1998.

Kauffman, S., *The Origins of Order: Self-Organization and Selection in Evolution*, Oxford University Press, Oxford, 1993.

Kauffman, S., What is life: was Schrodinger right?, in *The Next Fifty Years: Speculations on the Future of Biology*, pp. 83–114, M. P. Murphy and L. A. J. O'Neill, eds., Cambridge University Press, Cambridge, 1995. [Reprinted in Bedau and Cleland 2010]

Kauffman, S., *Investigations*, Oxford University Press, Oxford, 2000.

Kauffman, S., The emergence of autonomous agents, in *From Complexity to Life: On the Emergence of Life and Meaning*, pp. 47–71, N. H. Gregersen, ed., Oxford University Press, Oxford, 2003.

Kauffman, S., and P. Clayton, On emergence, agency, and organization, *Biol. Philos.*, **21**, 501–521, 2006.

Kellert, S. H., H. E. Longino, and C. K. Waters, eds., *Scientific Pluralism*, University of Minnesota Press, Minneapolis, MN, 2006.

Kim, J., Downward causation, in *Emergence or Reduction: Essays on the Prospects of Nonreductive Physicalism*, pp. 119–138, A. Beckermann, H. Flohr, and J. Kim, eds., Walter de Gruyter, New York, 1992.

Kim, J., Making sense of emergence, *Philos. Stud.*, **95**, 3–36, 1999.

Kirchman, D. L., *Processes in Microbial Ecology*, Oxford University Press, New York, 2012.

Kirschvink, J. L., Late Proterozoic low-latitude global glaciation: the snowball earth, in *The Proterozoic Biosphere: A Multidisciplinary Study*, pp. 51–52, J. W. Schopf and C. Klein, eds., Cambridge University Press, New York, 1992.

Kitcher, P., Species, *Philos. Sci.*, **51**, 308–333, 1984.

Klein, H. P., The Viking biological experiments on Mars, *Icarus*, **34**, 666–674, 1978a.

Klein, H. P., The Viking biology investigations: review and status, *Orig. Life Evol. Biosph.*, **9**, 157–160, 1978b.

Klein, H. P., The Viking biology experiments: epilogue and prologue, *Orig. Life Evol. Biosph.*, **21**, 255–261, 1991.

Knoll, A. H., The multiple origins of complex multicellularity, *Annu. Rev. Earth Planet. Sci.*, **39**, 217–239, 2011.

Knoll, A., M. J. Osborn, J. A. Baross, H. C. Berg, N. R. Pace, and M. Sogin, *Size Limits of Very Small Microorganisms: Proceedings of a Workshop*, National Academies Press, Washington, DC, 1999.

Knuuttila, T., and A. Loettgers, What are definitions of life good for? Transdisciplinary and other definitions in astrobiology, *Biol. Philos.*, **32**, 1185–1203, 2017.

Kolb, V., *Handbook of Astrobiology*, CRC Press, Taylor & Frances, Boca Raton, FL, 2018.

Koonin, E. V., The origin and early evolution of eukaryotes in the light of phylogenomics, *Genome Biol.*, **11**, 2–12, 2010.

Koonin, E. V., and W. Martin, On the origin of genomes and cells within inorganic compartments, *Trends Genet.*, **21**, 647–654, 2005.

Koonin, E. V., and P. Starokadomskyy, Are viruses alive? The replicator paradigm sheds decisive light on an old but misguided question, *Stud. Hist. Philos. Biol. Biomed. Sci.*, **59**, 125–134, 2016.

Koonin, E. N., and Y. I. Wolf, Is evolution Darwinian or/and Lamarckian?, *Biol. Direct*, **4**, 42, 2009.

Kopp, R. E., and J. L. Kirschvink, The identification and biogeochemical interpretation of fossil magnetotactic bacteria, *Earth Sci. Rev.*, **86**, 42–61, 2008.

Korzeniewski, B., Cybernetic formulation of the definition of life, *J. Theor. Biol.*, **209**, 275–286, 2001.

Kripke, S. A., *Naming and Necessity*, Harvard University Press, Cambridge, MA, 1972.

Kruger, K., P. J. Grabowski, A. J. Zaug, J. Sands, D. E. Gottshling, and T. R. Cech, Self splicing RNA: autoexcision and autocyclization of the ribosomal RNA intervening sequence of *Tetrahymena*, *Cell*, **31**, 147–157, 1982.

Kuhn, T. S., *The Structure of Scientific Revolutions*, University of Chicago Press, Chicago, IL, 1962. [2nd edition published 1970]

Kurland, C. G., L. J. Collins, and D. Penny, Genomics and the irreducible nature of eukaryote cells, *Science*, **312**, 1011–1014, 2006.

Ladyman, J., and D. Ross, with D. Spurrett, and J. Collier, *Everything Must Go: Metaphysics Naturalised*, Oxford University Press, Oxford, 2007.

Lane, N., and W. F. Martin, The origin of membrane bioenergetics, *Cell*, **151**, 1406–1416, 2012.

Lange, M., Life, 'artificial life,' and scientific explanation, *Philos. Sci.*, **63**, 225–244, 1995.

Langton, C. G., ed., *Artificial Life*, Addison Wesley, Redwood City, CA, 1989.

Langton, C. G., and C. Taylor, eds., *Artificial Life*, West View Press, Cambridge, MA, 2003.

La Scola, B., S. Audic, C. Robert, et al., A giant virus in amoebae, *Science*, **299**, 2033, 2003.

Lasne, J., A. Noblet, C. Szopa, et al., Oxidants at the surface of Mars: a review in light of recent exploration results, *Astrobiology*, **16**, 977–996, 2016.

Lavoisier, A. L., On the nature of water and on experiments which appear to prove that this substance is not strictly speaking an element but that it is susceptible of decomposition and recomposition, *Observations sur la Physique*, **23**, 452–455, 1783.

Lenski, R. E., M. R. Rose, S. C. Simpson, and S. C. Tadler, Long-term experimental evolution in *Escherichia coli*. I. Adaptation and divergence during 2,000 generations, *Am. Nat.*, **138**, 1315–1341, 1991.

Levere, T. H., *Transforming Matter: A History of Chemistry from Alchemy to the Buckyball*, Johns Hopkins University Press, Baltimore, MD, 2001.

Levin, G. V., The Viking labeled release experiment and life on Mars, *Instruments, Methods, and Missions for the Investigation of Extraterrestrial Microorganisms*, *Proc. SPIE*, **3111**, 146–161, 1997.

Levin, G. V., and P. A. Straat, Viking labeled release biology experiment: interim results, *Science*, **194**, 1322–1329, 1976.

Levin, G. V., and P. A. Straat, Recent results from the Viking labeled release experiment on Mars, *J. Geophys. Res.*, **82**, 4663–4667, 1977.

Levin, G. V., and P. A. Straat, Completion of the Viking labeled release experiment on Mars, *J. Mol. Evol.*, **14**, 167–183, 1979.

Levin, G. V., and Straat, P. A., The case for extant life on Mars and its possible detection by the Viking labeled release experiment, *Astrobiology*, **16**, 798–810, 2016.

Lewontin, R., Adaptation, in *The Dialectical Biologist*, pp. 65–84, R. Levins and R. Lewontin, eds., Harvard University Press, Cambridge, MA, 1985.

Lineweaver, C. H., We have not detected extraterrestrial life, or have we?, in *Life As We Know It*, pp. 445–457, J. Seckbach, ed., Springer, Dordrecht, 2006.

Linnaeus, C., *Systema Naturæ*, Laurentii Salvii, Stockholm, 1758.

Lipson, H., and Pollack, J. B., Automatic design and manufacture of robotic lifeforms, *Nature*, **406**, 974–978, 2000.

Liu, T., and W. S. Broecker, How fast does rock varnish grow?, *Geology*, **28**, 183–186, 2000.

Lloyd, E. A., *The Structure and Confirmation of Evolutionary Theory*, Greenwood Press, New York, 1988.

Lloyd, E. A., Holobionts as units of selection: holobionts as interactors, reproducers, and manifestors of adaptation, in *Landscapes of Collectivity in the Life Sciences*, pp. 351–368, S. B. Gissis, E. Lamm, and A. Shavit, eds., MIT Press, Cambridge, MA, 2018.

Locey, K. J., and J. T. Lennon, Scaling laws predict global microbial diversity, *Proc. Natl. Acad. Sci. USA*, **113**, 5970–5975, 2016.

Locke, J., *An Essay Concerning Human Understanding*, Oxford University Press, Oxford, 1689.

Lopez, D., H. Vlamakis, and R. Kolter, Biofilms, *Cold Spring Harb. Perspect. Biol.*, **2**, a000398, 2010.

Lovelock, J. E., Gaia as seen through the atmosphere, *Atmos. Environ.*, **6**, 579–580, 1972.

Lovelock, J. E., *The Ages of Gaia: A Biography of Our Living Earth*, Oxford University Press, Oxford, 2000.

Lovelock, J. E., and L. Margulis, At atmospheric homeostasis by and for the biosphere: the Gaia hypothesis, *Tellus*, **26**, 1–10, 1974.

Lowe, J. E., The causal autonomy of the mental, *Mind*, **102**, 629–644, 1993.

Luisi, P. L., The chemical implementation of autopoiesis, in *Self-Production of Supramolecular Structures: From Synthetic Structures to Models of Minimal Living Systems*, pp. 179–197, G. R. Fleischaker, C. Stephano, and P. L. Luisi, eds., Kluwer, Dordrecht, 1993.

Luisi, P. L., About various definitions of life, *Orig. Life Evol. Biosph.*, **28**, 613–622, 1998.

Luisi, P. L., Autopoiesis: a review and a reappraisal, *Naturwissenschaften*, **90**, 49–59, 2003.

Luisi, P. L., *The Emergence of Life: From Chemical Origins to Synthetic Biology*, Cambridge University Press, Cambridge, 2006.

Luisi, P. L., P. Walde, and T. Oberholzer, Lipid vesicles as possible intermediaries in the origin of life, *Curr. Opin. Colloid Interface Sci.*, **4**, 33–39, 1999.

MacCurdy, E., ed., *The Notebooks of Leonardo da Vinci (definitive edition in one volume)*, Konecky and Konecky, Old Saybrook, CT, 2003.

Machery, E., Why I stopped worrying about the definition of life... and why you should as well, *Synthese*, **185**, 145–164, 2012.

Machover, M., *Set Theory, Logic and Their Limitations*, Cambridge University Press, Cambridge, 1996.

Madigan, M.T., J. M. Martinko, et al., *Brock Biology of Microorganisms*, 11th edition, Prentice-Hall, Lebanon, 2006.

Magurran, A. E., and P. A. Henderson, Explaining the excess of rare species in natural species abundance distributions, *Nature*, **422**, 714–716, 2003.

Malyshev, D. A., K. Dhami, H. T. Quach, et al., Efficient and sequence-independent replication of DNA containing a third base pair establishes a functional six-letter genetic alphabet, *Proc. Natl. Acad. Sci. USA*, **109**, 12005–12010, 2012.

Malyshev, D. A., K. Dhami, T. Lavergne, et al., A semi-synthetic organism with an expanded genetic alphabet, *Nature*, **509**, 385–388, 2014.

Mann, A., Bashing holes in the tale of Earth's troubled youth, *Nature*, **553**, 393–395, 2018.

Margulis, L., *Symbiosis in Cell Evolution: Life and its Environment on the Early Earth*, W. H. Freeman, San Francisco, CA, 1981.

Margulis, L., and D. Sagan, *What is Life?*, University of California Press, Berkeley, CA, 1995.

Marshall, W. L., Hydrothermal synthesis of amino acids, *Geochim. Cosmochim. Acta*, **58**, 2099–2106, 1994.

Martin, W., and M. Russell, On the origins of cells: a hypothesis for evolutionary transitions from abiotic geochemistry to chemoautotrophic prokaryotes, and from prokaryotes to nucleated cells, *Philos. Trans. R. Soc. London, Ser. B*, **358**, 59–85, 2003.

Martin, W., J. Baross, D. Kelley, and M. J. Russell, Hydrothermal vents and the origin of life, *Nature Rev. Microbiol.*, **6**, 805–814, 2008.

Matthews, G., De Anima 2.2–4 and the meaning of life, in *Essays on Aristotle's De Anima*, pp. 185–193, M. C. Nussbaum and A. O. Rorty, eds., Oxford University Press, Oxford, 1992.

Mattingly, J., The structure of scientific theory change: models versus privileged formulations, *Philos. Sci.*, **72**, 365–389, 2005.

Maturana, H., and F. Varela, *Autopoiesis and Cognition: The Realization of the Living*, Reidel, Dordrecht, 1973.

Mayr, E., *Systematics and the Origin of Species from the Viewpoint of a Zoologist*, Columbia University Press, New York, 1942.

Mayr, E., The ontological status of species: scientific progress and philosophical terminology, *Biol. Philos.*, **2**, 145–166, 1987.

Mayr, E., *Toward a New Philosophy of Biology: Observations of an Evolutionist*, Harvard University Press, Cambridge, MA, 1988.

Mayr, E., The idea of teleology, *J. Hist. Ideas*, **53**, 117–135, 1992.

Mayr, E., *This is Biology: The Science of the Living World*, Harvard University Press, Cambridge, MA, 1997.

Mazzocchi, F., Images of thought and their relation to classification: the tree and the net, *Knowl. Org.*, **40**, 366–374, 2013.

McDaniel, L. D., E. Young, J. Delaney, F. Ruhnau, K. B. Ritchie, and J. H. Paul, High frequency of horizontal gene transfer in the oceans, *Science*, **330**, 50, 2010.

McKay, C. P., Origins of life, in *Van Nostrand Reinhold Encyclopedia of Planetary Sciences and Astrogeology*, J. Shirley and R. Fairbridge, eds., Van Nostrand, New York, 1994.

McKay, C. P., What is life – and how do we search for it in other worlds?, *PLoS Biol.*, **2**, e302, 2004.

McKay, C. P., Titan as the abode of life, *Life*, **6**, 1–15, 2016.

McKay, D. S., E. K. Gibson, K. L. Thomas-Keprta, et al., Search for past life on Mars: possible relic biogenic activity in Martian meteorite ALH84001, *Science*, **273**, 924–930, 1996.

McKie, R., 'Shadow biosphere' theory gaining scientific support, *The Observer*, 2013.

McKinsey, J. C. C., A. C. Sugar, and P. Suppes, Axiomatic foundations of classical particle mechanics, *J. Ration. Mech. Anal.*, **2**, 253–272, 1953.

McLaughlin, P., *Kant's Critique of Teleology in Biological Explanation*, Edwin Mellen Press, Lewiston, NY, 1990.

McLaughlin, B. P., The rise and fall of British emergentism, in *Emergence or Reduction? Essays on the Prospects of Nonreductive Physicalism*, pp. 49–93, A. Beckerman, H. Flohr, and J. Kim, eds., Walter de Gruyter, Berlin, 1992.

McMurry, J., and R. C. Fay, *Chemistry*, 3rd edition, Prentice-Hall, Englewood Cliffs, NJ, 2001.

Meierhenrich, U., *Amino Acids and the Asymmetry of Life*, Springer-Verlag, Berlin, 2008.

Mellor, C. H., Natural kinds, *Br. J. Philos. Sci.*, **28**, 299–312, 1977.

Mill, J. S., *A System of Logic*, Longmans Green, London, 1843. [1949 edition]

Miller, S. L., A production of amino acids under possible primitive Earth conditions, *Science*, **117**, 528–529, 1953.

Miller, S. L., Production of some organic compounds under possible primitive Earth conditions, *J. Am. Chem. Soc.*, **77**, 2351–2361, 1955.

Mitchell, S. D., Dimensions of scientific law, *Philos. Sci.*, **67**, 242–265, 2000.

Mitchell, S. D., Ceteris paribus – an inadequate representation for biological contingency, *Erkenntnis*, **57**, 329–350, 2002.

Mojzsis, S. J., G. Arrhenius, K. D. McKeegan, T. M. Harrison, A. P. Nutmann, and C. R. L. Friend, Evidence for life on Earth before 3,800 million years ago, *Nature*, **383**, 55–59, 1996.

Moreira, D., and P. López-García, Ten reasons to exclude viruses from the tree of life, *Nature Rev. Microbiol.*, **7**, 306–311, 2009.

Moreira, D., and P. López-García, Evolution of viruses and cells: do we need a fourth domain of life to explain the origin of eukaryotes?, *Philos. Trans. R. Soc. London, Ser. B.*, **370**, 2014.0327, 2015.

Morgan, C. L., *Instinct and Experience*, Macmillan, New York, 1912.

Morgan, M. S., and M. Morrison, Models as mediating instruments, in *Models as Mediators: Perspectives on Natural and Social Science*, pp. 10–37, M. S. Morgan and M. Morrison, eds., Cambridge University Press, Cambridge, 1999.

Morowitz, H. J., *Beginnings of Cellular Life: Metabolism Recapitulates Biogenesis*, Yale University Press, New Haven, CT, 1992.

Morowitz, H. J., B. Heinz, and D. W. Deamer, The chemical logic of a minimum protocell, *Orig. Life Evol. Biosph.*, **18**, 281–287, 1988.

Morrison, M., Models as autonomous agents, in *Models as Mediators: Perspectives on Natural and Social Science*, pp. 38–65, M. S. Morgan and M. Morrison, eds., Cambridge University Press, Cambridge, 1999.

Mueller, N. T., E. Bakacs, J. Combellick, Z. Grigoryan, and M. G. Dominguez-Bello, The infant microbiome development: mom matters, *Trends Mol. Med.*, **21**, 109–117, 2015.

Nasir, A., and G. Caetano-Anollés, A phylogenomic data-driven exploration of viral origins and evolution, *Sci. Adv.*, **1**, e1500527, 2015.

Navarro-Gonzáles, R., K. F. Navarro, J. de la Rosa, E. Iniguez, P. Molina, L. D. Miranda, P. Morales, E. Cienfuegos, P. Coll, F. Raulin, R. Amils, and C. P. McKay, The limitations on organic detection in Mars-like soils by thermal volatilization-gas chromatography-MS and their implications for the Viking results, *Proc. Natl. Acad. Sci. USA*, **103**, 16089–16094, 2006.

Nealson, K., and P. Conrad, Life: past, present, and future, *Philos. Trans. R. Soc. London, Ser. B*, **354**, 1923–1939, 1999.

Needham, P., The discovery that water is H_2O, *Int. Stud. Philos. Sci.*, **15**, 205–226, 2002.

Nelson, K. E., M. Levy, and S. L. Miller, Peptide nucleic acids rather than RNA may have been the first genetic molecule, *Proc. Natl. Acad. Sci. USA*, **97**, 3868–3871, 2000.

Newman, W. R., *Atoms and Alchemy*, University of Chicago Press, Chicago, IL, 2006.

Nielsen, P. E., and M. Egholm, An introduction to peptide nucleic acid, *Curr. Issues Mol. Biol.*, **1**, 89–104, 1999.

Noffke, N., D. Christian, D. Wacey, and R. M. Hazen, Microbially induced sedimentary structures recording an ancient ecosystem in the *ca.* 3.48 billion-year-old Dresser Formation, Pilbara, Western Australia, *Astrobiology*, **13**, 1103–1124, 2013.

NRC, *The Limits of Organic Life in Planetary Systems*, National Academy of Sciences, Washington, DC, 2007.

O'Malley, M. A., *Philosophy of Microbiology*, Cambridge University Press, Cambridge, 2014.

Oparin, A. I., *Life: Its Nature, Origin, and Development*, A. Synge, trans., Academic Press, New York, 1964. [Reprinted in Bedau and Cleland 2010]

Oren, A., Prokaryote diversity and taxonomy: current status and future challenges, *Philos. Trans. R. Soc. London, Ser. B*, **359**, 623–638, 2004.

Oren, A., Systematics of Archaea and Bacteria, in *Biological Science Fundamentals and Systematics*, volume II, pp. 162–186, A. Minelli and G. Contrfatto, eds., EOLSS, Oxford, 2009.

Orgel, L. E., Evolution of the genetic apparatus, *J. Mol. Biol.*, **38**, 381–393, 1968.

Orgel, L. E., The origin of life: a review of facts and speculation, *Trends Biochem. Sci.*, **23**, 491–495, 1998. [Reprinted in Bedau and Cleland 2010]

Orgel, L. E., Self-organizing biochemical cycles, *Proc. Natl. Acad. Sci. USA*, **97**, 12503–12507, 2000.

Orgel, L. E., Prebiotic chemistry and the origin of the RNA world, *Crit. Rev. Biochem. Mol. Biol.*, **39**, 99–123, 2004.

Orgel, L. E., The implausibility of metabolic cycles on the prebiotic Earth, *PLoS Biol.*, **6**, e18, 2008.

Oyama, V. I., B. J. Berdahl, and G. C. Carle, Preliminary findings of the Viking gas exchange experiment and a model for Martian surface chemistry, *Nature*, **265**, 110–114, 1977.

Pace, N. R., A molecular view of microbial diversity and the biosphere, *Science*, **276**, 734–740, 1997.

Pace, N., The universal nature of biochemistry, *Proc. Natl. Acad. Sci. USA*, **98**, 805–880, 2001. [Reprinted in Bedau and Cleland 2010]

Pályi, G., C. Zucchi, and L. Caglioti, Introduction: definitions of life, in *Fundamentals of Life*, pp. 2–13, G. Pályi, C. Zucchi, and L. Caglioti, eds., Elsevier, New York, 2002.

Perry, R. S., J. Dodsworth, J. T. Staley, and A. Gillespie, Molecular analysis of microbial communities in rock coatings and soils from Death Valley California, *Astrobiology*, **2**, 539, 2002.

Perry, R. S., B. Y. Lynne, M. A. Sephton, V. M. Kolb, C. C. Perry, and J. T. Staley, Baking black opal in the desert sun: the importance of silica in desert varnish, *Geology*, **34**, 537–540, 2006.

Philippe, N., M. Legendre, G. Doutre, et al., Pandoraviruses: amoeba viruses with genomes up to 2.5 Mb reaching that of parasitic eukaryotes, *Science*, **341**, 281–286, 2013.

Pinheiro, V. B., A. I. Taylor, C. Cozens, et al., Synthetic genetic polymers capable of heredity and evolution, *Science*, **336**, 341–344, 2012.

Pizzarello, S., and E. Shock, The organic composition of carbonaceous meteorites: the evolutionary story ahead of biochemistry, *Cold Spring Harb. Perspect. Biol.*, **2**, 1–19, 2010.

Pizzarello, S., G. W. Cooper, and G. J. Flynn, The nature and distribution of the organic material in carbonaceous chondrites and interplanetary dust particules, in *Meteorites and the Early Solar System II*, pp. 625–651, D. S. Laurett and H.Y. McSween, eds., University of Arizona Press, Tucson, AZ, 2006.

Planer, J. D., P. Yangqing, L. K. Andrew, L. V. Blanton, I. M. Ndao, P. I. Tarr, B. B. Warner, and J. I. Gordon, Development of the gut microbiota and mucosal IgA responses in twins and gnotobiotic mice, *Nature*, **534**, 263–266, 2016.

Popa, R., *Between Necessity and Probability: Searching for the Definition and Origin of Life*, Springer Verlag, Berlin, 2004.

Popper, K., *Conjectures and Refutations*, Routledge and Kegan Paul, London, 1963.

Powell, J. L., *Night Comes to the Cretaceous*, Harcourt and Brace, New York, 1998.

Powner, M. W., and J. D. Sutherland, Prebiotic chemistry: a new modus operandi, *Philos. Trans. R. Soc. London, Ser. B*, **366**, 2870–2877, 2011.

Powner, M. W., B. Gerland, and J. D. Sutherland, Synthesis of activated pyrimidine ribonucleotides in prebiotically plausible conditions, *Nature*, **459**, 239–242, 2009.

Psillos, S., *Scientific Realism: How Science Tracks Truth*, Routledge, London, 1999.

Putnam, H., What theories are not, in *Logic, Methodology, and the Philosophy of Science: Proceedings of the 1960 International Congress*, pp. 240–251, E. Nagel, P. Suppes, and A. Tarski, eds., Stanford University Press, Stanford, CA, 1962.

Putnam, H., Meaning and reference, *J. Philos.*, **70**, 699–711, 1973.

Putnam, H., The meaning of 'meaning', in *Language, Mind, and Knowledge: Minnesota Studies in the Philosophy of Science*, volume VII, pp. 131–193, K. Gunderson, ed., University of Minnesota Press, Minneapolis, MN, 1975. [Reprinted in Putnam 1979]

Putnam, H., ed., *Mind, Language and Reality*, pp. 215–272, Cambridge University Press, Cambridge, 1979.

Putnam, H., *Realism and Reason: Philosophical Papers*, volume 3, Cambridge University Press, Cambridge, 1983.

Quinn, R. C., H. F. H. Martucci, S. R. Miller, C. E. Bryson, F. J. Grunthaner, and P. J. Grunthaner, Perchlorate radiolysis on Mars and the origin of Martian soil reactivity, *Astrobiology*, **13**, 515–520, 2013.

Rasmussen, S., Aspects of information, life, reality, and physics, *Artif. Life*, **2**, 767–774, 1992. [Reprinted in Langton and Taylor 2003]

Rasmussen, S., L. Chen, M. Nilsson, and S. Abe, Bridging nonliving and living matter, *Artif. Life*, **9**, 269–316, 2003.

Ray, T. S., An approach to the synthesis of life, in *Artificial Life II, Santa Fe Institute Studies in the Sciences of Complexity Proceedings*, volume XI, pp. 371–408, C. G. Langton, C. Taylor, J. D. Farmer, and S. Rasmussen, eds., Addison-Wesley, Boston, MA, 1991.

Ray, T. S., Evolution, complexity, entropy, and artificial reality, *Physica D*, **75**, 239–263, 1994a.

Ray, T. S., An evolutionary approach to synthetic biology: zen and the art of creating life, *Artif. Life*, **1**, 179–209, 1994b. [Reprinted in Langton and Taylor 2003]

Ray, T. S., Artificial life programs and evolution, in *Companion to Evolution*, pp. 429–433, M. Ruse and J. Travis, eds., Harvard University Press, Cambridge, MA, 2009.

Redhead, M., The intelligibility of the universe, in *Philosophy of the New Millennium*, pp. 73–90, A. O'Hear, ed., Cambridge University Press, Cambridge, 2001.

Riskin, J., The naturalist and the emperor, a tragedy in three acts; or, how history fell out of favor as a way of knowing nature, *Know*, **2**, 85–110, 2018.

Roberts, G., *The Mirror of Alchemy: Alchemical Ideas in Images, Manuscripts and Books*, University of Toronto Press, Toronto, 1994.

Rokas, A., The origins of multicellularity and the early history of the genetic toolkit for animal development, *Annu. Rev. Genet.*, **42**, 235–251, 2008.

Rosen, R., *Life Itself*, Columbia University Press, New York, 1991.

Ross, D., and J. Ladyman, The alleged coupling-constitution fallacy and the mature sciences, in *The Extended Mind*, pp. 155–166, R. Menary, ed., MIT Press, Cambridge, MA, 2010.

Ruiz-Mirazo, K., J. Peretó, and A. Moreno, A universal definition of life: autonomy and open-ended evolution, *Orig. Life Evol. Biosph.*, **34**, 323–346, 2004. [Reprinted in Bedau and Cleland 2010]

Ruse, M., *Charles Darwin*, Blackwell, Malden, 2008.

Russell, M. J., and A. J. Hall, The emergence of life from iron monosulphide bubbles at a submarine hydrothermal redox and pH front, *J. Geol. Soc. London*, **154**, 377–402, 1997.

Ryder, G., Mass flux in the ancient Earth–Moon system and benign implications for the origin of life on Earth, *J. Geophys. Res.*, **107**, 5022, 2002.

Sagan, C., The definition of life, in *Encyclopaedia Britannica*, 14th edition, 1970. [Reprinted in Bedau and Cleland 2010]

Sagan, L., On the origin of mitosing cells, *J. Theor. Biol.*, **14**, 225–274, 1967.

Sapp, J., *Genesis: The Evolution of Biology*, Oxford University Press, Oxford, 2003.

Sapp, J., The prokaryote–eukaryote dichotomy: meanings and mythology, *Microbiol. Mol. Biol. Rev.*, **69**, 292–305, 2005.

Schaffner, J., Monism, in *The Stanford Encyclopedia of Philosophy*, E. N. Zalta, ed., 2007. (https://plato.stanford.edu/archives/win2018/entries/monism)

Schilthuizen, M., and R. Stouthamer, Horizontal transmission of parthenogenesis-inducing microbes in Trichogramma wasps, *Proc. Biol. Sci.*, **264**, 361–366, 1997.

Schmitt-Kopplin, P., Z. Gabelico, R. D. Gougeon, et al., High molecular diversity of extraterrestrial organic matter in Murchison meteorite revealed 40 years after its fall, *Proc. Natl. Acad. Sci. USA*, **107**, 2763–2768, 2010.

Schneider, E. D., and J. J. Kay, Life as a manifestation of the second law of thermodynamics, *Math. Comput. Model.*, **19**, 25–48, 1994.

Schopf, J. W., A. B. Kudryavtsev, A. D. Czaja, and A. B. Tripathi, Evidence of Archean life: stromatolites and microfossils, *Precambrian Res.*, **158**, 141–155, 2007.

Schrödinger, E., *What is Life?*, Cambridge University Press, Cambridge, 1944.

Schuerger, A. C., and B. C. Clark, Viking biology experiments: lessons learned and the role of ecology in future Mars life-detection experiments, *Space Sci. Rev.*, **135**, 233–243, 2008.

Schulze-Makuch, D., and W. Bains, The first cell and the origin of life challenge, in *The Cosmic Zoo: Complex Life on Many Worlds*, pp. 35–52, Springer, Berlin, 2017.

Schulze-Makuch, D., and L. N. Irwin, The prospect of alien life in exotic forms in other worlds, *Naturwissenschaften*, **93**, 155–172, 2006.

Schulze-Makuch, D., and L. N. Irwin, *Life in the Universe: Expectations and Constraints*, 2nd edition, Springer Verlag, Berlin, 2008.

Schulze-Makuch, D., J. N. Head, J. M. Houtkooper, et al., The biological oxidant and life detection (BOLD) mission: a proposal for a mission to Mars, *Planet. Space Sci.*, **67**, 57–69, 2012.

Seager, S., and W. Bains, The search for signs of life on exoplanets at the interface of chemistry and planetary science, *Sci. Adv.*, **1**, e1500047, 2015.

Seager, S., W. Bains, and J. J. Pelowski, Towards a list of molecules as potential biosignature gasses for the search for life on exoplanets and applications to terrestrial biochemistry, *Astrobiology*, **16**, 465–485, 2016.

Sefah, K., Z. Yang, K. M. Bradley, S. Hoshika, E. Jiménez, L. Zhang, G. Zhu, S. Shanker, F. Yu, D. Turek, W. Tan, and S. A. Benner, In vitro selection with artificial expended genetic information systems, *Proc. Natl. Acad. Sci. USA*, **111**, 1449–1454, 2014.

Segré, D., D. Ben-Eli, and D. Lancet, Compositional genomes: prebiotic information transfer in mutually catalytic noncovalent assemblies, *Proc. Natl. Acad. Sci. USA*, **97**, 4112–4117, 2000.

Segré, D., D. Ben-Eli, D. W. Deamer, and D. Lancet, The lipid world, *Org. Life Evol. Biosph.*, **31**, 119–145, 2001.

Sender, R., S. Fuchs, and R. Milo, Revised estimates for the number of human and bacterial cells in the body, *PLoS Biol.*, **14**, e1002533, 2016.

Shanahan, T., *The Evolution of Darwinism: Selection, Adaptation, and Progress in Evolutionary Biology*, Cambridge University Press, New York, 2004.

Shapiro, J. A., Thinking about bacterial populations as multicellular organisms, *Annu. Rev. Microbiol.*, **52**, 81–104, 1998.

Shapiro, R., A replicator was not involved in the origin of Life, *IUBMB Life*, **49**, 173–176, 2000.

Shapiro, R., Small molecule interactions were central to the origin of life, *Q. Rev. Biol.*, **81**, 105–126, 2006. [Reprinted in Bedau and Cleland 2010]

Shen, B., L. Dong, S. Xiao, and M. Kowalewksi, The Avalon explosion: evolution of Ediacara morphospace, *Science*, **319**, 81–84, 2008.

Sheng, G.-P., H.-Q. Yu, and X.-Y. Li, Extracellular polymeric substances (EPS) of microbial aggregates in biological wastewater treatment systems: a review, *Biotechnol. Adv.*, **28**, 882–894, 2010.

Shields, C., *Order in Multiplicity: Homonymy in the Philosophy of Aristotle*, Oxford University Press, Oxford, 2002.

Shields, C., *Aristotle's De Anima*, Clarendon Press, Oxford, 2013. [Reprinted in Bedau and Cleland 2010]

Simon, H. A., The axiomatization of physical theories, *Philos. Sci.*, **37**, 16–26, 1970.

Singh, R. K., H. Chang, D. Yan, K. M. Lee, D. Ucmak, K. Wong, M. Abrouk, B. Farahnik, M. Nakamura, T. H. Zhu, T. Bhutani, and W. Liao, Influence of diet on the gut microbiome and implications for human health, *J. Transl. Med.*, **15**, 73–88, 2017.

Smith, E., and H. Morowitz, Universality in intermediary metabolism, *Proc. Natl. Acad. Sci. USA*, **101**, 13168–13173, 2004.

Smith, E., and H. Morowitz, *The Origin and Nature of Life on Earth: The Emergence of the Fourth Geosphere*, Cambridge University Press, Cambridge, 2015.

Smolen, L., *The Life of the Cosmos*, Oxford University Press, Oxford, 1997.

Sneed, J., *The Logical Structure of Mathematical Physics*, Reidel, Dordrecht, 1971.

Sober, E., Learning from strong functionalism: prospects for strong artificial life, in *Artificial Life II, Santa Fe Institute Studies in the Sciences of Complexity Proceedings*, C. G. Langton, C. Taylor, J. D. Farmer, and S. Rasmussen, eds., Addison-Wesley, Boston, MA, 1991. [Reprinted in Bedau and Cleland 2010]

Sogin, M. L., H. G. Morrison, J. A. Huber, et al., Microbial diversity in the deep sea and the underexplored 'rare biosphere', *Proc. Natl. Acad. Sci. USA*, **103**, 12115–12120, 2006.

Soros, L. B., and K. O. Stanley, Identifying necessary conditions for open-ended evolution through the artificial life world of chromaria. In *Artificial Life 14: Proceedings of the Fourteenth International Conference on the Synthesis and Simulation of Living Systems*, pp. 793–800, H. Sayama, J. Rieffel, S. Risi, R. Doursat, and H. Lipson, eds., MIT Press, Cambridge, MA, 2014.

Sosa Torres, M. E., J. P. Saucedo-Vázquez, and P. M. H. Kroneck, The magic of dioxygen, in *Sustaining Life on Planet Earth: Metalloenzymes Mastering Dioxygen and Other Chewy Gases, Metal Ions in Life Sciences*, volume 15, pp. 1–12, P. Kroneck and M. Sosa Torres, eds., Springer, Cham, 2015.

Stamati, K., V. Mudera, and U. Cheema, Evolution of oxygen utilization in multicellular organisms and implications for cell signaling in tissue engineering, *J. Tissue Eng.*, **2**, 2041731411432365, 2011.

Standish, R., Open-ended artificial evolution, *Int. J. Comput. Intell. Appl.*, **3**, 167–175, 2003.

Stanford, P. K., *Exceeding Our Grasp: Science, History, and the Problem of Unconceived Alternatives*, Oxford University Press, Oxford, 2006.

Steele, A., F. M. McCubbin, and M. D. Fries, The provenance, formation, and implications of reduced carbon phases in Martian meteorites, *Meteorit. Planet. Sci.*, **51**, 2203–2225, 2016.

Stubbendieck, R. M., C. Vargas-Bautista, and P. D. Straight, Bacterial communities: interactions to scale, *Front. Microbiol.*, **7**, 1234, 2016.

Summers, Z. M., J. A. Gralnick, and D. R. Bond, Cultivation of an obligate Fe(II)-oxidizing lithoautotrophic bacterium using electrodes, *mBio*, **4**, e00420–12, 2013.

Suppe, F., *The Structure of Scientific Theories*, University of Illinois Press, Chicago, IL, 1977.

Suppe, F., *The Semantic Conception of Scientific Theories and Scientific Realism*, University of Illinois Press, Chicago, IL, 1989.

Suppes, P., *Introduction to Logic*, Van Nostrand, New York, 1957.

Suppes, P., What is scientific theory?, in *Philosophy of Science Today*, pp. 55–67, S. Morgenbesser, ed., Basic Books, New York, 1967.

Suppes, P., *Representation and Invariance of Scientific Structures*, CSLI Publications, Stanford, CA, 2002.

Sure, S., M. L. Ackland, A. A. J. Torriero, A. Adholeya, and M. Kochar, Microbial nanowires: an electrifying tale, *Microbiology*, **162**, 2017–2028, 2016.

Suttle, C. A., Marine viruses – major players in the global ecosystem, *Nature Rev. Microbiol.*, **5**, 801–812, 2007.

Szostak, J. E., D. P. Bartel, and P. L. Luisi, Synthesizing life, *Nature*, **409**, 387–390, 2001.

Tarski, A., The semantic conception of truth, *Philos. Phenom. Res.*, **4**, 341–376, 1944.

Tarski, A., The concept of truth in formalized languages, in *Logic, Semantics and Metamathematics*, Oxford University Press, New York, 1951.

Tegmark, M., Is 'the theory of everything' merely the ultimate ensemble theory?, *Ann. Phys.*, **270**, 1–51, 1998.

Tegmark, M., *Our Mathematical Universe: My Quest for the Ultimate Nature of Reality*, Alfred A. Knopf, New York, 2014.

Teller, P., Twilight of the perfect model, *Erkenntnis*, **55**, 393–415, 2001.

ten Kate, I. L., Organics on Mars?, *Astrobiology*, **10**, 589–603, 2010.

Tessera, M., Origin of evolution versus origin of life: a shift of paradigm, *Int. J. Mol. Sci.*, **12**, 3445–3458, 2011.

Theis, K. R., A. Venkataraman, J. A. Dycus, K. D. Koonter, E. N. Schmitt-Matzen, A. P. Wagner, K. E. Holekamp, and T. M. Schmidt, Symbiotic bacteria appear to mediate hyena social odors, *Proc. Natl. Acad. Sci. USA*, **110**, 19832–19837, 2013.

Thomas-Keprta, K. L., D. A. Bazylinski, J. L. Kirschvink, et al., Elongated prismatic magnetite crystals in ALH84001 carbonate globules: potential Martian magnetofossils, *Geochim. Cosmochim. Acta*, **64**, 4049–4081, 2000.

Tian, R. M., L. Cai, W. P. Zhang, H. L. Cao, and P. Y. Qian, Rare events of intragenus and intraspecies horizontal transfer of the 16S rRNA gene, *Genome Biol. Evol.*, **7**, 2310–2320, 2015.

Timmis, J. N., M. A. Ayliffe, C. Y. Huang, and W. Martin, Endosymbiotic gene transfer: organelle genomes forge eukaryotic chromosomes, *Nature Rev. Genet.*, **5**, 123–135, 2004.

Trifonov, E. N., Vocabulary of definitions of life suggests a definition, *J. Biomol. Struct. Dynamics*, **29**, 259–266, 2011.

Tsapin, A. I., M. G. Goldfeld, G. D. McDonald, et al., Iron (VI): hypothetical candidate for the Martian oxidant, *Icarus*, **147**, 68–78, 2000.

Tsokolov, S. A., Why is the definition of life so elusive?, *Epistemol. Consid. Astrobiol.*, **9**, 401–412, 2009.

Tyrell, T., *On Gaia: A Critical Investigation of the Relationship between Life and Earth*, Princeton University Press, Princeton, NJ, 2013.

Urbano, P., and F. Urbano, Nanobacteria: facts or fancies?, *PLoS Pathog.*, **3**, e55, 2007.

Uwins, P. J. R., R. Webb, and A. P. Taylor, Novel nano-organisms from Australian sandstones, *Am. Mineral.*, **83**, 1541–1550, 1998.

Vago, J. L., F. Westall, A. J. Coates, and O. I. Korablev, Habitability on early Mars and the search for biosignatures with the ExoMars rover, *Astrobiology*, **17**, 471–519, 2017.

Van Fraassen, B., On the extension of Beth's semantics to physical theories, *Philos. Sci.*, **37**, 325–339, 1970.

Van Fraassen, B., *The Scientific Image*, Oxford University Press, Oxford, 1980.

Van Fraassen, B., *Laws and Symmetry*, Oxford University Press, Oxford, 1989.

van Regenmortel, M. H. V., The metaphor that viruses are living is alive and well, but it is not more than a metaphor, *Stud. Hist. Philos. Biol. Biomed. Sci.*, **59**, 117–124, 2016.

Vasas, V., E. Szathmáry, and M. N. Santos, Lack of evolvability in self-sustaining autocatalytic networks constraints metabolism-first scenarios for the origin of life, *Proc. Natl. Acad. Sci. USA*, **107**, 1470–1475, 2010.

Vella, J., *Aristotle: A Guide for the Perplexed*, Continuum International, London, 2008.

Vetsigian, K., C. Woese, and N. Goldenfield, Collective evolution and the genetic code, *Proc. Natl. Acad. Sci. USA*, **103**, 10696–10701, 2006.

von Neumann, J., *Mathematical Foundations of Quantum Mechanics*, R. T. Beyer, trans., Princeton University Press, Princeton, NJ, 1996 (original published in 1932).

Wacey, D., M. R. Kilburn, M. Saunders, J. Cliff, and M. D. Brasier, Microfossils of sulphur metabolizing cells in 3.4-billion-year-old rocks of Western Australia, *Nature Geosci.*, **4**, 698–702, 2011.

Wächtershäuser, G., Before enzymes and templates: theory of surface metabolism, *Microbiol. Rev.*, **52**, 452–484, 1988.

Wächtershäuser, G., The case for the chemoautotrophic origin of life in an iron-sulfide world, *Orig. Life Evol. Biosph.*, **20**, 173–176, 1990.

Wächterhäuser, G., Groundworks for an evolutionary biochemistry: the iron-sulfur world, *Prog. Biophys. Mol. Biol.*, **58**, 85–201, 1992.

Wang, B., M. Yao, L. Lv, Z. Ling, and L. Li, The human microbiota in health and disease, *Eng. Microecol. Rev.*, **3**, 71–82, 2017.

Ward, P. D., and D. Brownlee, *Rare Earth: Why Complex Life Is Uncommon in the Universe*, Springer Verlag, New York, 2000.

Watamabe, K., and K. Nishio, Electric power from rice paddy fields, in *Paths to Sustainable Energy*, pp. 563–580, A. Ng, ed., INTECH, Rijeka, 2010.

Weinbauer, M. G., Ecology of prokaryote viruses, *FEMS Microbiol. Rev.*, **28**, 127–181, 2004.

Weisberg, M., Three kinds of idealization, *J. Philos.*, **104**, 639–659, 2007.

Weisberg, M., *Simulation and Similarity: Using Models to Understand the World*, Oxford University Press, Oxford, 2013.

Wessner, D., Discovery of the giant mimivirus, *Nature Educ.*, **3**, 61, 2010.

Westall, F., Morphological biosignatures in early terrestrial and extraterrestrial materials, *Space Sci. Rev.*, **135**, 95–114, 2008.

Westfall, R., *The Construction of Modern Science*, Cambridge University Press, Cambridge, 1977.

Whittaker, E. T., *A History of the Theories of the Aether and Electricity: From the Age of Descartes to the Close of the Nineteenth Century*, Longmans, Green and Co., London, 1910.

Whittaker, R. H., New concepts of kingdoms of organisms, *Science*, **163**, 150–160, 1969.

Wieland, W., Aristotle's *Physics* and the problem of inquiry into principles, in *Articles on Aristotle*, volume 1, pp. 127–140, J. Barnes, M. Schofield, and R. Sorabji, eds., Duckworth, London, 1975.

Williams, M. B., Deducing the consequences of evolution: a mathematical model, *J. Theor. Biol.*, **29**, 343–385, 1970.

Williams, T. A., P. G. Foster, C. J. Cox, and T. M. Embley, An archaeal origin of eukaryotes supports only two primary domains of life, *Nature*, **504**, 231–236, 2013.

Wimsatt, W., False models as means to truer theories, in *Neutral Models in Biology*, pp. 23–55, M. H. Nitecki and A. Hoffman, eds., Oxford University Press, Oxford, 1987.

Wimsatt, W. C., *Re-engineering Philosophy for Limited Beings: Piecewise Approximations to Reality*, Harvard University Press, Cambridge, MA, 2007.

Winther, R. G., The structure of scientific theories, in *The Stanford Encyclopedia of Philosophy*, E. N. Zalta, ed., 2015.

Woese, C. R., *The Genetic Code: The Molecular Basis for Gene Expression*, Harper and Row, New York, 1967.

Woese, C., The universal ancestor, *Proc. Natl. Acad. Sci. USA*, **95**, 6854–6859, 1998.

Woese C. R., The archaeal concept and the world it lives in: a retrospective, *Photosynth. Res.*, **80**, 371–372, 2004.

Woese, C. R., and G. Fox, Phylogenetic structure of the prokaryotic domain: the primary kingdoms, *Proc. Natl. Acad. Sci. USA*, **74**, 5088–5090, 1977.

Woese, C., L. J. Magrum, and G. E. Fox, Archaebacteria, *J. Mol. Evol.*, **11**, 245–252, 1978.

Woese, C. R., O. Kandler, and M. L. Wheelis, Towards a natural system of organisms: proposal for the domains Archaea, Bacteria, and Eukarya, *Proc. Natl. Acad. Sci. USA*, **87**, 4576–4579, 1990.

Wu, D. Y., M. Wu, A. Halpern, et al., Stalking the fourth domain in metagenomic data: searching for, discovering, and interpreting novel, deep branches in marker gene phylogentic trees, *PLoS One*, **6**, e18011, 2011.

Xie, J., and P. G. Schultz, Adding amino acids to the genetic repertoire, *Curr. Opin. Chem. Biol.*, **9**, 548–554, 2005.

Yamada, C., A. Gotoh, M. Sakanaka, M. Hattie, K. A. Stubbs, A. Katayama-Ikegami, J. Hirose, S. Kurihara, T. Arakawa, M. Kitaoka, S. Okuda, T. Katayama, and S. Fushinobu, Molecular insight into evolution of symbiosis between breast-fed infants and a member of the human gut microbiome *Bifidobacterium longum*, *Cell Chem. Biol.*, **24**, 515–524, 2017.

Yang, Z., F. Chen, J. B. Alvarado, and S. A. Benner, Amplification, mutation, and sequencing of a six letter synthetic genetic system, *J. Am. Chem. Soc.*, **133**, 15105–15112, 2011.

Yarus, M., *Life from an RNA World: The Ancestor Within*, Harvard University Press, Cambridge, MA, 2011.

Yooseph, S., G. Sutton, D. B. Rusch, et al., The Sorcerer II global ocean sampling expedition: expanding the universe of protein families, *PLoS Biol.*, **5**, e77, 2007.

Yue, J., X. Hu, H. Sun, Y. Yang, and J. Huang, Widespread impact of horizontal gene transfer on plant colonization of land, *Nature Commun.*, **3**, 1152, 2012.

Zeleny, M., ed., *Autopoiesis: A Theory of Living Organization*, North-Holland, New York, 1981.

Zemach, E. M., Putnam's theory on the reference of substance terms, *J. Philos.*, **73**, 116–127, 1976.

Zent, A. P., and C. P. McKay, The chemical reactivity of the Martian soil and implications for future missions, *Icarus*, **108**, 146–157, 1994.

Zilber-Rosenberg, I., and E. Rosenberg, Role of microorganisms in the evolution of animals and plants: the hologenome theory of evolution, *FEMS Microbiol. Rev.*, **32**, 723–735, 2008.

Zuckerkandl, E., and L. Pauling, Molecules as documents of evolutionary history, *J. Theor. Biol.*, **8**, 357–366, 1965.

Zykov, V., E. Mytilinaios, B. Adams, and H. Lipsone, Self-reproducing machines, *Nature*, **435**, 163–164, 2005.

Index

abstraction, 69–70, 163, 165–166, 185–186
    Goldilocks level of, 90–93, 102–103
adenosine triphosphate (ATP), 156–157
Allan Hills meteorite (ALH84001), 183–184, 190, 214
amino acids, 107–112, 197–198
    chirality, 108
    nonstandard, 111
anomalies, 173, 179–181, 184–185, 191–194, 211–213, *See also* Kuhn, Thomas, scientific paradigms
Archaea, *See* Domains of life
Aristotle, 8–17, 24, 85–86, 118
    on motion, 13, 85–86
    on souls, 12–14
    on the nature of life
        self-nutrition, 8, 10–15
        self-reproduction, 8, 10–15, 24
    teleology, *See* causation
    theory of material substance, 51
    theory of terrestrial matter, 91
artificial life (ALife), 40, 161–171
    strong hard ALife, 164–166
    strong soft ALife, 161–164
    wet ALife, 167–170
autocatalytic reaction networks, 26–27
autopoiesis, 22–23, 162–164

Bacteria, *See* Domains of life
banded iron formations, 190
Bedau, Mark, 42, 63, 77–78, 161
Benner, Steve, 43, 63
big data movement, 96–97
biological individual, 143–154
    biological autonomy, 144–146
    Darwinian individual, 144–145, 148–149, 152
    metabolic individual, 144
Boden, Margaret, 163

carbon, 113–115, 202
carbon dioxide, 39, 127, 174–175
causal-descriptive theories, 59
causal theory of reference, 58
causation, 12, 15
    goal-directed self-causation, 9, 11–13, 15–16, 18–21, 27, 30, 145
Cech, Thomas
    discovery of ribozymes, 178–180
Chang, Hasok, 86, 98–99, 101, *See also* Phlogiston
Chao, Lin, 47
computer metaphor, 20–21
contingencies, 90–91, 120–123, 130–131
Copernican revolution, 85–86
correspondence rules, 66–67
Crick, Francis, 200
    central dogma of molecular biology, 179
cultivation, 207–208

Darwin, Charles
    biological individuality, 144–145, 204
    Darwinian algorithm, 162–164
    theory of evolution, 18–21, 23–24, 40–43, 95, 133–134, 139–140
deductively closed axiom systems, 64–68, 72, 78
definitions
    function of, 64, 78, 80
    ideal, 34, 48–50, 52, 73
    lexical, 46–47
    nonstandard, 63–64, 73–79
    operational, 47–48, 80
    stipulative, 52–54, 57
    traditional, 63
definitions of life
    evolutionary, 23–24, 40
        chemical Darwinian, 23–24, 41–43, 60, 122–123, 126, 133
        Darwinian, 41

definitions of life (cont.)
  metabolic
    autopoietic, 22–23, 43–44, 49, 122
    chemical-metabolic, 22, 24, 37–40, 123
    thermodynamic, 22, 36–37
Descartes, René, 15, 86
descriptive theory, 53–59
  cluster theory, 58–59
desert varnish, 212–213
disunity, 102, 132, 218
Domains of life, 119
  Archaea, 116–117, 119–120, 134–135, 181–182,
    205–206
  Bacteria, 95–96, 115–117, 119–120, 134–135,
    140–141, 146–147, 155–157, 167–168,
    173–174, 181–183, 205–206, *See also*
    species concepts
  Eukarya, 105, 115–120, 134–135, 142, 152–153,
    203
  prokaryote–eukaryote distinction, 115, 118–119, 181
Doolittle, Ford, 154
Duhem, Pierre, 98, 101
Dupré, John, 98
Dyson, Freeman, 122

Einstein, Albert
  theory of relativity, 95, 97
electron transport chains (ETCs), 155–158
emergence, 28, 32
emergentism, 16–18, 27
end-Cretaceous mass extinction, 120–121
Eukarya, *See* Domains of life
exoplanets, 190–191

Gánti, Tibor, 43–44
Gibbs, Josiah, 92
Great Oxygenation Event, 117, 142

Haeckel, Ernst, 118
Herschel, William
  discovery of Uranus, 177–178
horizontal gene transfer (HGT), 95, 133–138,
    140–143, 202–204
Hume, David, 17, 30

idealization, 69, 90
inertia, 87
interpretation, 70–71
isomorphism, 71–72

Kant, Immanuel, 15–16
Kauffman, Stuart, 27, 37, 43, 124, 128
Kepler, Johannes
  on inertia, 86
Kingdoms of life, 118–119
Kuhn, Thomas
  scientific paradigms, 54–55, 101, 176–178

Lamarckian evolution, 139–143
Langton, Chris, 161
last universal common ancestor (LUCA), 135, *See also*
    $N = 1$ problem
lateral gene transfer, *See* horizontal gene transfer
Lavoisier, Antoine, 25, 50–51, 92
Lego Principle, 188–189
life
  alternative origin of, 196–198
  defining criteria, 173–174, 186–187, 193
  functional characteristics of
    genetic-based reproduction, 8–9, 14–15, 128–129
    self-organization, 8, 14–15, 129
  nature of, 127–130
  tentative criteria, 176, 184–194
Linnaeus, Carl, 118
Lipson, Hod, 165
Locke, John, 53–54
logical positivism, 65
logicomathematical structures, 72–73, 88
luminiferous aether, 94–95

Mars, *See also* Viking missions to Mars
  life-detection missions, 180–181
mass, 86–88
mathematical theories, 64, 66–67, 71–72, 78
Mayr, Ernst, 136
metabolism, 122, 155–158, 172–173, 186–187, *See*
    also origins of life, metabolism-first
  anabolism, 155–156
  catabolism, 155
metagenomic methods, 208–211
microbes, 47, 105–106, 146–155, 201–215, *See also*
    Domains of life
  alien, 47, 168, 172–176, 180–181, 187, 205–208,
    210–211
  biofilms, 149–153
  holobionts, 148–149
  rock-powered ecosystems, 156–158
microscopy, 206–207
models, 68–73, 89, 128–129
  standard model, 71–72
modern evolutionary synthesis, 20
monism, 99–102, 129, *See also* pluralism
mutation, 140–142

$N = 1$ problem, 105–114, 117, 161, 169–170
nanobacteria, 214–215
nanobes, 213–215
natural kinds, 4, 34–35, 53–60, 64, 66–67, 73–74, 129
necessary and sufficient conditions, 4–5, 63–64, 67,
    75–80
Newton, Isaac
  theory of motion, 3, 5, 15–16, 54–55, 65–66, 70, 77,
    83, 85–88, 90–91, 97, 143
nucleic acids, 107, 109–113, 125–126, 167, 198–200
  XNA, 112

On the Origin of Species, See Darwin, Charles
ontology, 83–85, 87–88, 96–97
  alternative ontologies, 93, 101
  inadequate ontologies, 87
  premature commitment to, 93
Orgel, Leslie, 128
origins of life, 33, 114–127, 135, 145, 158–159
  genes-first, 25, 122, 124, 128–129
  Lipid World, 124, 128
  metabolism-first, 26, 122–123, 128–129,
    198
  RNA World, 27–31, 106, 122, 124–129, 199–201,
    203
  pre-RNA World, 28, 125
  Small Molecule (SM) World, 26–29, 106, 122–124,
    126–128, 201

paradigm, *See* Kuhn, Thomas, scientific paradigms
persistence, 153–154
phlogiston, 94, 98
pluralism, 91, 96, 98–102, 129, 132, 160
  promiscuous realism, 98
  Scientific Pluralism, 98, 100–102
  species pluralism, 137–138
polymerase chain reaction (PCR), *See* metagenomic
    methods
proteins, 107–109, 113, 159, 167, 179–180,
    197–199
Putnam, Hilary, 76
  Twin-Earth thought experiment, 55–59

quantum nonlocality, 18

Ray, Thomas, 40
  Tierra, 162–163
regeneration, 152–154
replicas, 72

Sagan, Carl, 36–37
Saturn, 177–178
  life on Titan, 106, 113, 189

scientific theories, 4–6, 63–81, 89
  exceptions to, 100
  pragmatic conception of, 79
  semantic conception of, 63, 68–73, 77–80, 84, 89
  syntactic conception of, 63–68, 84
set theoretic structures, 69–71
shadow biosphere, 195, 201–216
silicon, 38, 113
species concepts, 95, 135–139
  bacterial species problem, 96, 136–138
  biological species concept (BSC), 137
  phylogenetic species concept (PSC), 137

theoretical definitions, 72–73, 77–80
theoretical framework, *See* ontology
theoretical identity statements, 52–54, 57, 73–76
theoretical kinds, 66
theoretical principles, 84
theoretical terms, 66–67
theory ladenness of observation, *See* Kuhn, Thomas,
    scientific paradigms
tree of life, 135–136, 159

unity, 101–102

Venter, Craig
  Synthia 3.0, 167–168
Viking missions to Mars, 38–40, 172–176, 180
  gas chromatograph mass spectrometer (GCMS), 39,
    174–175, 210
  labeled release experiment (LR), 39, 172–176, 180,
    186–187, 192
  perchlorate, 175
viruses, 145–146
  giant viruses, 182–183
vitalism, 15–18

water, 50–53, 56, 60–61, 74–76
Weismann, August, 139–140
Whittaker, R. H., 119
Woese, Carl, 42, 181, 203–204